D1598023

OLIVER
70
OLIVER

INTERCONTINENTAL
26
Owned By: Ed Spiess
Restored By: Marshall
& Solomon
1
ARMSTRONG

The TRACTOR FACTOR

THE WORLD'S RAREST CLASSIC FARM TRACTORS

ROBERT N. PRIPPS

PHOTOGRAPHS BY
RALPH W. SANDERS
AND ANDREW MORLAND

First published in 2015 by Voyageur Press,
an imprint of Quarto Publishing Group USA Inc.,
400 First Avenue North, Suite 400, Minneapolis, MN 55401 USA.
Telephone: (612) 344-8100 Fax: (612) 344-8692

quartoknows.com
Visit our blogs at quartoknows.com

10 9 8 7 6 5 4 3 2 1

ISBN: 978-0-7603-4893-2

Library of Congress Cataloging-in-Publication Data

Pripps, Robert N., 1932– author.
The tractor factor : the world's rarest classic farm tractors / Robert N. Pripps ; photographs by Ralph W. Sanders and Andrew Morland.
pages cm
ISBN 978-0-7603-4893-2 (hardback)
1. Antique and classic tractors. 2. Farm tractors. I. Sanders, Ralph W., 1933– II. Morland, Andrew, 1947– III. Title.
TL233.25.P755 2015
629.225'2—dc23

2015018779

Acquiring Editor: Todd R. Berger
Project Manager: Caitlin Fultz
Art Director: Cindy Samargia Laun
Cover Design: Matt Simmons
Book Design and Layout: Wendy Holdman

On the front cover: *Ralph W. Sanders*
On the back cover: *Ralph W. Sanders*
On the frontis: 1936 Oliver 70. The six-cylinder engine was designed to run on 70-octane gasoline. *Ralph W. Sanders*
On the title page: 1948 Intercontinental C-26. A four-cylinder Continental engine of 28 horsepower powered the C-26. *Ralph W. Sanders*
On the contents page: 1929 Case L. The L was one of the most powerful tractors of its day. Its four-cylinder engine produced 47 belt horsepower. *Ralph W. Sanders*

Printed in China

CONTENTS

Introduction

When it comes to classic farm tractors, it seems every interested person has an opinion of the brand and model that should be at the top of the list. There are those who have an interest in collectible tractors, and then there are those who are *really* into collectible tractors. Hopefully, all will find something of interest in this book, and some may find a new avenue to explore.

We have selected about 100 of what my editor and I thought to be the rarest and most interesting tractors, but still not so rare that the few in existence are already locked in the sheds of collectors. Some that we've selected are, admittedly, not all that rare, but from a brand that sold a hundred thousand or more per year, a model from the same manufacturer selling a mere 10,000 might still be worth a look.

One chapter includes tractors that signaled turning points in the industry, such as the first one to use a turbocharger. Another lists tractors with unique features, such as low-production orchard and industrial versions. We also included a chapter on the smallest tractors. For most hobbyists, small means light and easier to haul. It also means less storage space and often a lower-cost restoration.

Finally, pictures: we obtained photos from famous photographers Andrew Morland and Ralph Sanders. We used pictures from a broad cross section of tractor brands, including those of non-US makers. We also included several working scale-model tractors, as there is a lot of interest in these machines.

The fun in tractor collecting comes from attending the shows. I've had good times taking my very common 2N Ford Ferguson to shows (probably a thousand in my county alone). But something even slightly rare will usually draw a crowd.

SAMSON
MODEL-M

Fordson

MASSEY HARRIS

SENIOR Tractors

TRACTORS IN THIS CATEGORY—those made before or during the early 1920s—are among the most fascinating simply because of their age. While many exist, they represent a general level of difficulty in restoration and presentation that restricts them to only the more serious collectors. Any in this age group would be considered rare vintage farm tractors.

1910 Pioneer 30
1913 Rumely Oil-Pull 30-60
1913 Holt 60
1914 Allis-Chalmers 10-18
1918 International Harvester Motor Cultivator
1919 International 8-16
1919 Samson M
1919 Avery 14/28
1919 Case Crossmotor 10/18
1919 Moline Universal D
1921 Renault HO
1921 Eagle 16/30 H

1910 PIONEER 30

The most famous product of the Pioneer Tractor Manufacturing Company of Winona, Minnesota, the Pioneer 30 was a truly impressive machine. Just consider the fact that it was first built in 1910 and the level of technology employed at that time. The Pioneer 30 was over 20 feet long and 8 feet wide. It weighed 23,600 pounds. Its drive wheels were 8 feet in diameter. Collecting tractor examples such as this requires true dedication. Need parts? You make them from scratch. To haul it, you need a semi. To store it, you need a barn.

The Pioneer company was incorporated in 1909 to produce what it called its "Field Motor." It advertised its 30-horsepower (drawbar) tractor for $100 per horsepower, or for $3,000—a princely sum at that time. The belt-pulley horsepower rating was 60. The Pioneer 30 was sold both in the United States and in Canada. Canadian machines were manufactured in Calgary, Alberta. Pioneer made more than 1,000 between 1910 and 1927. The company also made smaller and, unbelievably, a larger version, but none of these seem to have survived.

The Pioneer 30 employed a four-cylinder horizontally opposed transverse-mounted engine of 1,232 ci. A two-speed transmission provided a road speed of 6 mph. It was rated for 10 14-inch plow bottoms. The tractor was equipped with an impulse magneto, a front-mounted radiator, a shaft-driven fan, and a water pump.

Other interesting features of the Pioneer 30 were its standard enclosed cab, possibly the first in the industry; optional headlights for night work; and a sprung front axle.

1910 Pioneer 30. An impressive machine was the Pioneer 30 with 8-foot drive wheels and 5-foot front wheels. Production ran from 1910 to 1915. The tractor was 20 feet long and weighed more than 23,000 pounds. *Ralph W. Sanders*

1913 RUMELY OIL-PULL 30-60

1913 Rumely Oil-Pull 30-60. The Type E was rated at 16 drawbar and 30 belt horsepower. The "Oil-Pull" name signified that the tractor used oil as a coolant. *Ralph W. Sanders*

Manufactured by the Advance-Rumely Company of La Porte, Indiana, the Oil-Pull 30-60 was another impressive, gigantic machine with a two-cylinder engine of 1,885 ci and a weight of 26,000 pounds.

Rumely started out as a blacksmith shop in 1852. Meinrad Rumely and his brother, John, who were German immigrants, began building threshers and steam engines. The first Oil-Pull tractor was built and tested in 1909, and the first example was sold in 1910. The company acquired the Advance Thresher Company in 1911 and changed its name to Advance Rumely Thresher Company. In 1931 Allis-Chalmers bought out Advance Rumely.

The 30-60 Oil-Pull, also known as the Type E, came out in 1911 and remained in production through 1923. It was tested at the University of Nebraska test facility in Lincoln (Test No. 8) in 1920. Its performance amazed the test engineers, as its slow-turning (375 rpm) engine produced a drawbar pull of 10,025 pounds. It could have been rated much higher than its claimed 30-60 drawbar/belt horsepower, as it demonstrated 50 drawbar and 76 belt horsepower.

A characteristic of the Oil-Pull line was a cooling tower in which exhaust gases induced air flow through the radiator to cool the oil and to cool the engine. Oil, which had a much higher boiling temperature than water, worked much better with kerosene fuel. The Oil-Pulls consumed water in almost equal parts to fuel, however, as water injection was used during hard pulls to keep cylinder head temperatures under control.

The Type E Rumely 30-60 used chain and bolster pivot-axle steering and was equipped with a pneumatic start system. The transmission offered one speed forward and one in reverse. Top and working speed was 2 mph.

Advance Rumely built the Type E from 1911 to 1923. It had a sterling reputation for reliability and pulling power. The two-cylinder transverse engine with a bore and stroke of 10 x 12 inches operated at only 375 rpm. Allis-Chalmers acquired Advance Rumely in 1931. *Ralph W. Sanders*

1913 HOLT 60

Charles Holt moved from New Hampshire to San Francisco in 1864. He made the journey by ship, sailing around Cape Horn. Once in San Francisco, he set up a carriage wheel company importing hardwoods from his family's lumber business back in New Hampshire. Trouble developed when the wooden wheel spokes shrunk in the dry California climate. Holt then moved his operation to Stockton, California, which had extremely low humidity, and there dried his lumber for a full year before fashioning it into wheels. At this time, in 1883, the company was known as The Stockton Wheel Company.

At the same time, great farms developed in the San Joaquin Valley near Stockton, which then became the center for agricultural equipment production. Holt first ventured into threshers and then combines, but upward of 40 horses were required to power the combines, so going to steam power was the natural solution. Now, the combines became too heavy for the spongy peat soil, so tracks were added. This led to tracked, or crawler tractors, first with steam power, but soon with gasoline engines. The first was the Holt 60 in 1911.

For its time, the Holt 60 was a remarkable machine. The 1913 version pictured here sold for

$4,200. It weighed 40,000 pounds and had a 1,230-ci overhead-valve, four-cylinder engine. Large band clutches in the rear allowed de-clutching each track individually to allow for steering, which was then accomplished through a "tiller" wheel on the front of the machine. The exposed master clutch and flywheel rotated just to the left of the driver's feet. For starting the engine, holes were provided in the flywheel whereby a crowbar could be inserted and used for leverage to rotate the flywheel through a compression stroke. The machine had a drawbar rating of 50 horsepower and 60 on the belt pulley.

Steering was accomplished then by the tiller wheel on the nose of the tractor. Tiller wheel tractors were made by Holt until 1921. In 1925, Holt and Best combined to form Caterpillar Corporation. *Ralph W. Sanders*

1913 Holt 60. This tractor was popular with large-acreage farmers in both California and the Great Plains of the United States and Canada. It had an open engine flywheel with a starting hole, in which a crowbar was inserted to roll the flywheel through a compression stroke of the big four-cylinder engine. The big drums on the rear contained individual clutches released for steering. *Ralph W. Sanders*

1914 ALLIS-CHALMERS 10-18

The Allis-Chalmers 10-18 (10 horsepower on the drawbar, 18 on the belt pulley) was powered by a two-cylinder, horizontally opposed engine mounted transverse, or cross-mounted, with the clutch, gearing, and belt pulley on the left side. It was a three-wheel design, popular at the time, with a single front wheel in line with the right drive wheel. This allowed the front and right rear wheel to run in the plow furrow for easier maintenance of the line of draft. The smaller upper cylindrical tank was for gasoline for starting; the lower tank was for kerosene, the normal operating fuel. The 303-ci engine was equipped with a centrifugal governor and operated at 720 rpm. A high-tension impulse magneto was fitted. The transmission gave only one speed forward and one in reverse. Top speed was just under 3 mph. Although the 10-18 looked light, it tipped the scales at 4,800 pounds. Extant pictures taken in 1915 show the 10-18 pulling a plow with three 14-inch bottoms.

The 10-18 was also known simply as the "Farm Tractor" in company advertising. It was built in A-C's West Allis, Wisconsin, factory. It was their third attempt at a tractor design; the first was a rototiller outfit, and the second was a tracked (crawler) vehicle called the Tractor-Truck. These, along with the 10-18, had short terms of production because of low sales. Allis-Chalmers went on to become a major player in the American tractor industry.

1914 Allis-Chalmers 10-18. Known simply as the "Farm Tractor," this three-wheel design was A-C's third design, although the first two did not see production. The 4,800-pound machine was capable of pulling three 14-inch plows. The engine had two cylinders; the transmission had only one speed and reverse. The 10-18 boasted a one-piece steel frame that "would not sag." About 2,700 were made. *Ralph W. Sanders*

1918 INTERNATIONAL HARVESTER MOTOR CULTIVATOR

The development of motor-driven crop cultivators began as early as 1915 at International Harvester. By 1917 fewer than 500 copies of the design were built and many of those were sold to farmers on a trial basis. Most versions had the engine mounted over dual rear-drive wheels; the whole package swiveled for steering. These experimental machines were hand built and subject to constant revision. Both a LeRoi engine of 138 ci and a Continental of only 67 ci were used in the experiments. The fact that the engine was mounted so high on the rig made it top-heavy and subject to rollovers when turning. This, plus a high price, made it a hard sell. Farmers generally could not afford a tractor and a Motor Cultivator, and this fact led to the combination known as the Farmall.

Of the several configurations tried, the type pictured held the most promise. Crop visibility was superb with the cultivator shovels clearly visible in front of the operator. Further, foot pedals allowed shifting of the shovels when they got too close to the plants.

Both the LeRoi and the Continental engines were four-cylinder types that operated on kerosene. Both produced 10 to 12 horsepower. No transmission was used; the engine-steering assembly was turned around for reverse. The rig weighed about 4,000 pounds.

No transmission was provided, just a clutch. For reverse, the entire engine-drive-wheel assembly was turned around. Crop visibility for the driver was not obstructed in any way. Rear-wheel steering quickly moved the shovels away from the plants, plus there were foot pedals to further dart the shovels away if they got too close. *Ralph W. Sanders*

1918 International Harvester Motor Cultivator. Starting in 1916, the International Harvester Motor Cultivator was offered to farmers, but its utility was limited to cultivating row crops. While the machine was very good at that, it proved to be too expensive for its limited use. *Ralph W. Sanders*

1919 INTERNATIONAL 8-16

In a stark break from the past, International Harvester introduced this smaller, lighter tractor in 1917, replacing the two-cylinder 8-16 Mogul. Mogul tractors were sold by McCormick dealers in those days while Deering dealers had a comparable but different Titan line. With the International 8-16, both the McCormick and Deering names were dropped and the so-called competing dealerships were eliminated. Production continued through 1922, with many experimental variations tried along the way, including a full crawler version. Initially, the price was $1,150, but with competition from Henry Ford's Fordson in the 1922 tractor price war, the price dropped to $670 with a two-bottom Little Wonder plow thrown in for good measure.

The entire design of the International 8-16 was taken from the International Type G truck. It used the same overhead-valve, four-cylinder engine with the radiator and fan in back over the flywheel housing, and the same hood sheet metal. The engine displaced 284 ci and operated at 1,000 rpm. It was equipped with a Dixie magneto and an Ensign carburetor. A three-speed transmission gave speed of 2, 3, and 4 mph. Final drive was by roller chain. It weighed in for its University of Nebraska Tractor Test (No. 25) at 3,360 pounds. International built more than 33,000 Model 8-16s in the Chicago Tractor Works plant, where they also built the line of Mogul tractors.

1919 International 8-16. The drawbar rating was 8 horsepower, with 16 horsepower available from the belt pulley. *Ralph W. Sanders*

KEROSENE
TRACTOR
TIONAL
HP

1919 SAMSON M

1919 Samson M. The 1919 Samson M was General Motors' answer to the popular Fordson by Ford. It cost more but had standard features, such as a belt pulley, fenders, and a governor, that cost extra on the Fordson. The 276-ci engine was larger than that of the Fordson. It was also heavier but had only a two-speed transmission. *Ralph W. Sanders*

Determined to meet and beat Henry Ford in the tractor business, William Durant, chairman of General Motors Corporation, bought out the Samson Tractor Works of Stockton, California. He also bought a plant in Janesville, Wisconsin, and moved the Samson outfit there, which was organized as the Janesville Machine Company. In December 1918, the company announced the Samson M, with a selling price of $650. This was $30 more than the Fordson at that time, but fenders, a belt pulley, and a governor were standard equipment, all of which cost extra on the Fordson. In appearance, the tractors looked much the same, except the Samson had a distinctive six-inch gap between the fuel tank and the radiator water tank. It was thought that this elongation of the tractor, moving the front wheels, radiator, and so on forward, gave the tractor better balance and avoided the rearing problem of the Fordson. Also, the Samson at 3,300 pounds was 600 pounds heavier than the Fordson.

The M engine was a four-cylinder, L-head type of 276 ci versus 251 ci for the Fordson. Operating speed was 1,100 rpm. It used a multiple-disk wet clutch and a Simms magneto. A two-speed transmission was provided, while the Fordson offered three speeds. The University of Nebraska Tractor Test data indicated the Samson bested the Fordson by less than 1 horsepower on the belt and by about 1 horsepower on the drawbar. Both tractors were built on the "unit-frame" concept where the engine, transmission, and rear-axle housings served as the tractor's frame.

General Motors offered several other models in its Samson venture and while the M did fairly well, Samson experienced huge losses overall and left the tractor business in 1922. By then the Fordson was selling for around $390 and was the largest-selling tractor brand in the world.

1919 AVERY 14/28

Not to be confused with the B. F. Avery Company of Louisville, Kentucky, this Avery Company was founded in Galesburg, Illinois, in 1874, manufacturing miscellaneous farm tools but moved to Peoria in 1884. The company first got into steam-powered tractors, but internal-combustion engine-driven machines followed as early as 1909. These early experiments did not meet with success, however, until 1912 when the company came out with a line of two- and four-cylinder, horizontally opposed–engine tractors. These, while not sophisticated in styling or fancy mechanisms, did garner a reputation for dependability. The 14/28 (drawbar/belt horsepower rating) was tested at the University of Nebraska (Test No. 42) in 1920.

The 14/28, built from 1919 to 1924, was the smallest four-cylinder version built. Like all Avery tractors in the line, an exhaust-induced cooling tower coolant radiator was used. Also, peculiar to Avery tractors of the time, the engine, radiator, and fuel tank slid back and forth to change between its two forward gears. The tractor weighed about 7,500 pounds, could pull up to a four-bottom plow, and could travel at 3.5 mph in high gear. The overhead-valve engine displaced 470 ci.

The Avery Company of Peoria, Illinois, endured several reorganizations but finally went out of existence at the start of World War II.

1919 Avery 14/28. Avery offered eight models of this tractor in 1919, with the 14/28 in the middle of the pack. The market was changing, however, and these big heavyweights were falling out of favor. *Ralph W. Sanders*

1919 CASE CROSSMOTOR 10/18

The Case Model Crossmotor Model 10/18 came out in 1918 as a replacement for the 9-18B. It was much the same, but the engine speed was increased by 150 rpm. Another feature of the 10/18 was the Sylphon System, a thermosyphon cooling system that included a water pump. The 10/18 used an internally expanding clutch, and the Case-patented air-wash intake filtration system was carried over from the 9-18. Production continued through 1921, when it was replaced by the 12/20 Crossmotor. Because of their more interesting color, the green-and-blue Case tractors like the 10/18 are more collectable than the later, all-gray models. Pinstriping, especially of the fenders, added interest.

The transverse engine was an overhead-valve vertical four-cylinder type displacing 236 ci. It produced a maximum of 18 horsepower at 1,050 rpm on kerosene fuel. A two-speed transmission was installed, and high gear gave a maximum speed of 3.5 mph. The machine weighed 3,760 pounds and was rated for two 14-inch plows. The width of the Crossmotor design made these Case tractors wider than most of this era, limiting their use for row-crop purposes.

1919 Case Crossmotor 10/18. This tractor was Case Threshing Machine Company's answer to the competition from the Fordson. *Ralph W. Sanders*

1919 MOLINE UNIVERSAL D

Manufactured by the Moline Plow Company, the D was built from 1918 to 1923. It rated 9 horsepower on the drawbar and 18 on the belt pulley and was actually a two-wheel machine that required a sulky, or a trailed implement which provided the back wheels and a seat for the driver. The Universal was probably the first attempt by any manufacturer to make an all-purpose tractor. Some versions had water tanks in each wheel for added weight; others simply had cast cement in the wheels. The Universal had a Remy electric governor/generator combination and a self-starter. Steering was mechanical by articulation. The engine was a four-cylinder L-head, vertical inline type with 192 cid. It only had a single forward gear and reverse. Top speed was 4 mph. The Universal weighed in at about 3,600 pounds with ballast.

1919 Moline Universal D. To capitalize on the growing demand for the all-purpose tractor, Moline Plow Company came out with the Universal concept, in which plows, cultivators, mowers, and the like were attached to the engine unit in front. *Ralph W. Sanders*

1921 RENAULT HO

Senior tractor collectors are broadening their search areas looking for remarkable and unusual items to add interest to their collections by going to foreign makes. Europe is a fertile field (no pun intended) for tractors rarely seen in America. Renault of Paris, France, has produced some unusual and very interesting machines.

Renault got into automobiles as early as 1898, and then into farm tractors in 1910 with a crawler design. This, of course, led to tanks for World War I, for which several types were made. The first, the FT-1, made for the French Army, was a very light design with a sloping hood over the front engine and with the radiator tilted forward at about a 45-degree angle. This reduced its height and protected it behind the engine to reduce its vulnerability. Their first postwar farm tractor, the H1, was also a crawler. The HO was a four-wheel version of the H1 crawler and had the same type of engine/hood/radiator/gas tank combination, which also provided fairly good driver visibility. The fact that the gas tank bore a tank logo was a reminder to the farmer of the tractor's wartime origins.

The HO used a four-cylinder L-head engine that produced 20 horsepower at 1,600 rpm. A three-speed transmission drove through epicyclic gears in the rear wheels. The HO was produced from 1920 to 1926.

In 2003, Claas Company bought Renault Agriculture from the French government, which had nationalized the firm in 1945.

1921 Renault HO. Built by the French automobile company, the Model HO used a four-cylinder L-head engine that produced 20 horsepower at 1,600 rpm. The radiator was positioned behind the engine similar to that of the International 8-16. Renault produced the HO from 1920 to 1926. *Andrew Morland*

1921 EAGLE 16/30 H

Eagle Manufacturing Company of Appleton, Wisconsin, was a tractor force to be reckoned with from before World War I until the start of World War II. Its first tractor, a 30-horsepower, two-cylinder machine, was produced in 1906. A similar model appeared in 1911. The two-cylinder 16/30 H came out in 1916 and was produced until 1930, when the company abruptly switched to six-cylinder tractors. Very few of the two-cylinder varieties survive. The Eagle Company did not resume production after World War II.

The Eagle H used a two-cylinder, horizontally opposed, overhead-valve engine of 804 ci, mounted with the crankshaft transverse to the direction of travel. Rated engine speed was 500 rpm. The Waterloo Boy lookalike weighed 7,200 pounds and had a belt-driven cooling fan. A two-speed transmission gave speeds of 2 and 3 mph. Eagle, at this time, made its own engines. The tractor used automotive-type steering, but a similar 12/22 F, made at the same time, used chain and bolster swing axle steering.

1921 Eagle 16-30 H. The tractor used automotive-type steering, a belt-driven fan, and an internally expanding clutch. *Michael Geldorf*

The LUMINARIES

THE PRACTICE OF AGRICULTURE CHANGED LITTLE from the beginning of time to the middle of the nineteenth century. Then changes came rapidly. The invention of the reaper in 1831 by Cyrus McCormick opened the floodgates. The reaper created the need for the thresher, which created the need for the steam engine, and so on. The deluge of Model T cars from Henry Ford's factory led to the Good Roads Movement, a national effort that worked with local governments to organize and coordinate road building, and what the Good Roads Movement needed was tractor power for road construction. Farmers also benefited from both the harvesting and tillage inventions and the Model T, which gave them a newfound freedom. Low-cost (compared to the steamers) tractors ensued, and familiarity with internal combustion engines became commonplace. Some of these tractors have a special place in the history of agriculture and are quite rare simply because there were not so many of them in the first place. The American Fordson is not included, even though it established the conventional configuration of the tractor for the next 10 years, because there were so many of them built—more than 850,000. There are still so many around that while they are interesting, they are hardly rare.

1924 FARMALL REGULAR

Competition from the Fordson caused International Harvester to lose its first place in tractor sales. At the same time, Harvester had been working on a motorized corn cultivator with only marginal success. The idea to combine the several tractor functions with the cultivator led to a real breakthrough in power farming—the Farmall. Its high stance, narrow front wheels, rear power takeoff, belt pulley, and individual turning brakes gave the Farmall the ability to finally replace all of the horses on the farm.

The Farmall has the distinction of being the first row-crop tractor to be tested at the University of Nebraska (Test No. 117). It was so successful in identifying the row-crop configuration that almost all other tractor makers followed suit. The Farmall was powered by a four-cylinder, vertical, inline overhead-valve engine of 221 ci. A three-speed transmission gave forward speeds of 2, 3, and 4 mph. It weighed 4,100 pounds and was capable of pulling a two-bottom plow. The fuel was kerosene.

For corn cultivation, the high stance and the narrow front wheels (tricycle configuration) allowed the rear axle to ride above the crop until it was quite mature, while the front wheels ran between two rows. Thus two rows could be cultivated at once. At the row end, cable-operated steering brakes allowed the Farmall to spin around and go into the next two rows.

These first Farmalls are rare today because not many survived from their production from 1924 to 1931. In 1932, an improved version called the F-20 came out. The previous versions acquired the moniker "Regular" for the purpose of differentiating between the two.

1924 Farmall Regular. The first Farmalls are especially collectable because not many survived from their production period from 1924 to 1931. They can be readily identified by their open steering gear on the nose and by a tall air intake pipe, generally with a mesh balloon-type air cleaner. *Randy Leffingwell*

1924 WATERLOO BOY N

This second iteration of John Deere's successful Waterloo Boy tractor was the first tractor to successfully complete the University of Nebraska Tractor Test series (Test No. 1, 1920). (The first iteration, the R, will be covered in Chapter 6.)

The historic Nebraska Tractor Law was introduced in the Nebraska State Legislature in 1919 by Representative Wilmot F. Crozier, himself a farmer who had been "burned" by a series of oversold, underperforming tractors. Crozier's bill stated that no tractor could be sold in the state of Nebraska without a permit that certified that the tractor had met published performance claims, and that such testing would be carried out by the University of Nebraska. The law is still in effect and has become the standard of the world. (See www.nebraskatractortest.edu for test reports.)

The Waterloo Boy N was a worthy test article and one of the best tractors of its time. It was introduced in 1918 and produced through 1924, when it was supplanted by the John Deere D. It can be said that the Waterloo Boy got John Deere into the tractor business.

The N was powered by a two-cylinder, side-by-side, horizontal, transverse engine of 465 ci. This type of engine would characterize Deere tractors until 1960. The N also offered a two-speed transmission and automotive-type steering. It was rated at 25 belt horsepower and 12.5 on the drawbar. Only around 20,000 were built, which, considering their age, makes them quite rare.

1924 Waterloo Boy N. John Deere bought the Waterloo Gasoline Engine Company of Waterloo, Iowa, in 1918 and got into the tractor business overnight. The N was powered by a two-cylinder, side-by-side, horizontal, transverse engine of 465 ci. *Ralph W. Sanders*

JOHN
DEERE
HILL
KEROSENE
WATERLOO BOY

1926 CATERPILLAR 2 TON

Weighing in at 4,040 pounds, the Caterpillar 2 Ton was aptly named. When submitted for testing at the University of Nebraska (Test No. 86, 1922) the papers with it said it had formerly been known as the T-35. This tractor is historically significant because it was one of three brought into the merger by Holt in the Holt-Best merger that formed the Caterpillar Tractor Company in 1925. Holt had registered the name Caterpillar back in 1910 and had called its tractors "Holt Caterpillars." The other two models brought in by Holt were the 5 Ton and the 10 Ton, both of which were soon dropped in favor of the Best contributions to the union. Nevertheless, the 2 Ton found favor with both farmers and loggers because of its power, traction, and manageable size.

Some interesting features of the 2 Ton included an overhead cam shaft engine, T-handle steering, clutch actuators operating oil-cooled clutches, and a master brake pedal on the left side of the cockpit with dual steering brake pedals on the right side. It managed a maximum drawbar pull of 81 percent of its own weight during its University of Nebraska test. The engine was a 251-ci four-cylinder vertical inline type, driving through a three-speed transmission uniquely mounted behind the rear axle. Top speed was a sporty 5.5 mph. The 2 Ton was in production from 1921 to 1928. As with people, it gained weight as it got older; it weighed almost 5,000 pounds by the end of production. Its list price in 1927 was $1,850.

1926 Caterpillar 2 Ton. The four-cylinder engine of the Cat 2 Ton had a unique overhead camshaft. The 2 Ton also had T-handle steering clutch control and a master foot clutch for the driver's left foot. The steering brakes were together on the right side of the platform. *Ralph W. Sanders*

1929 IRISH FORDSON

A food crisis loomed in Great Britain during the early stages of World War I, due in large part to the attacks on shipping by German U-boats. The British Ministry of Munitions mounted an all-out plowing campaign to increase the acreage devoted to the growing of grains for food. Henry Ford was invited to send over 6,000 of his newly developed Fordson tractors. Ford agreed to not only send the first of his factory's production to Britain, he also offered to establish a factory in Cork, Ireland, and to make a gift of all patents and drawings. Problems of many kinds prevented the production of Irish Fordsons until the first one rolled off the assembly line on July 4, 1919. Then political problems developed in 1922 when the British government imposed restrictions on industrial products from the recently created Irish Free State. So production was curtailed in Cork until it was revived again in 1929.

The 1929 Irish Fordson was an entirely new machine. Now labeled the Fordson N, it replaced the original, which was called the Fordson F. Cork production began with serial number 757369 and ended in 1932 at serial number 779135. Thus there were fewer than 28,000 N Irish tractors produced, accounting for their special interest to tractor buffs.

Changes to the Fordson N engine included an increase in the cylinder bore, from 4 inches to 4.125 inches, to increase the displacement from 251 ci to 267 ci; a high compression head; a gasoline carburetor; a water pump; a high-tension impulse magneto; a new heavy front axle; and cast, heavy front wheels. These changes corrected most of the problems inherent in the Fordson F and greatly improved its performance.

1929 Irish Fordson. The venerable Fordson was in need of an upgrade when production was transferred to Ireland for the second time, in 1928. A high-tension magneto was added, as was a water pump. The engine displacement increased and heavy cast front wheels helped the balance problem. An optional electrical system became available with self-starting and lights. *Ralph W. Sanders*

1929 CASE L

The 1929 Case L is a historically significant tractor because it was the pivotal tractor in Case history that made the switch from the cross-motor design to the now-conventional configuration. Also, only 6,000 Ls were made in 1929 and 1930. Further, the L continued in production until it was replaced by the LA. During that time, only about 34,000 were built.

Following the 1918 Fordson's lead, most tractor companies switched their product lines to four-cylinder inline vertical engines with car-like hoods. Leon R. Clausen had left his position as vice president at Deere & Company in 1924 and became president of Case. He immediately stopped work on the cross-motors and directed the development of the L, a design in the Fordson's

image. The big difference between the two was the Fordson used a power-robbing worm drive while the Case L used a new and unique dual roller chain drive.

The Case L was also almost twice as heavy as the Fordson, weighing 5,300 pounds versus the Fordson at 2,700 pounds. Further, the L had an engine displacement of 403 ci against 251 ci of the Fordson. The Case was rated for three 14-inch plows and the Fordson only two. Both had three-speed transmissions.

For competition, the L was more comparable to Deere's D, which had just been introduced at the time Clausen moved to Case. Both were rated 30-horsepower drawbar and 40 on the belt. Both sold for around $1,300 in 1929–30.

1929 Case L. The L was a conventional standard-tread tractor with the engine crank parallel to the line of travel. A hand clutch was used, while final drive was by a roller chain. The L featured a high-speed (1,100-rpm) four-cylinder engine. *Ralph W. Sanders*

1930 ALLIS-CHALMERS U

The Allis-Chalmers U has the distinction of being the first farm tractor to be offered for sale with rubber tires. The first 7400 U model (and its row-crop running mate, the UC) had a four-cylinder Continental L-head engine, but subsequent tractors used Allis' own four-cylinder OHV engine. By 1930, the pictured U had the Allis engine, which displaced 300 ci, operated on either kerosene or gasoline, and produced 33 belt horsepower. The U weighed about 5,100 pounds.

Production of the U ran from 1929 to 1941 with rubber tires as standard equipment from 1932 on. The U and subsequent A-C tractors were painted Persian Orange, which became an Allis-Chalmers trademark.

Allis-Chalmers touted the use of rubber tires by sponsoring the U in speed trials and races at county fairs. One vain-yet-famous race driver, Barney Oldfield, had it in his contract that he would always win. Another famous driver of the time was Ab Jenkins, who attained 67 mph driving a U on the Bonneville Salt Flats in Utah. The top speed of tractors sold to the public, however, was a little over 10 mph.

1930 Allis-Chalmers U. Not very fast on steel wheels, but the U did break speed records on rubber tires. The tractor was originally designed for the United Tractor Company, but after it ceased operations, A-C began casting their own name in the radiator header. *Ralph W. Sanders*

1931 EAGLE 6A

Eagle Manufacturing Company, a division of the Four Wheel Drive Auto Company of Appleton, Wisconsin, made tractors as early as 1906. Early on, its line was the traditional Waterloo Boy–type heavyweights with two-cylinder engines. With the Eagle 6, the company jumped all the way from two cylinders to six. Besides the Eagle 6A, standard-tread plowing tractor with 40-horsepower engine, there was an Eagle 6B Universal (all-purpose) and an Eagle 6C (standard-tread using the same engine as the 6B). The models 6B and C were designed to compete with the current crop of 10 to 20 horsepower–rated tractors such as the Fordson and Case CC. The 6A was built from 1930 to 1937.

The engine for the Eagle tractor line was the Hercules six-cylinder L-head vertical inline with 339 cid. Using gasoline, it produced 40 belt horsepower at 1,400 rpm. In top gear of the three-speed transmission, the 6A could make about 5 mph. Electric starter and lights were optional. The Eagle 6A weighed in at 4,650 pounds.

All of the Eagle tractors are historically significant because of the very early tractor production in Wisconsin. After World War II, Eagle did not resume operations.

1931 Eagle 6A. Rated at 37 horsepower on the belt pulley and 22 on the drawbar, the Eagle 6A could pull a four-bottom plow. It was powered by a 339-ci Hercules six-cylinder engine. *Ralph W. Sanders*

EAGLE-6

1934 JOHN DEERE A

The John Deere A goes down in history as the first tractor to employ a hydraulic implement lift. It was also the first to use splined rear axles to facilitate wheel spacing adjustments. Deere management bet the company on the A. They had garnered a good reputation with their Waterloo Boy and their D, but this was the depth of the Depression, which hit farmers especially hard, and the GP, Deere's first row-crop model, had not met with much success. Nevertheless, the company took the plunge and never looked back. The A was a smashing success.

Other features of the A were a centerline rear PTO; individual steering brakes with pedals on either side of the platform; a transverse-mounted, two-cylinder, horizontal engine of 309 ci; a one-piece differential housing; a four-speed transmission; and a thermosyphon cooling system (no water pump required). The tractor could be fitted with a front-mounted two-row cultivator or a two-bottom (14-inch) plow. A belt pulley, driven directly through the clutch from the engine crankshaft, was standard equipment.

At first, only the row-crop version was offered, but in 1935, wide front wheels (AW) and a single front wheel (AN) options were added. Still later, high-clearance (Hi-Crop) versions were offered as well.

The most significant collectability factor for the John Deere A is that the first of the series, up to Serial Number 414809 (approximately 4,800 tractors), were built without a covering tube over the shaft that drives the fan. These are now affectionately known to collectors as "Open Fan Shaft As."

1934 John Deere A. Deere's A was rated to pull two 16-inch plows. The model was built from 1934 to 1952 in row-crop, standard-tread, orchard, and industrial configurations. *Ralph W. Sanders*

JOHN DEERE
JOHN DEERE
GENERAL PURPOSE

1935 JOHN DEERE B

The John Deere B was introduced in the depths of the Great Depression in 1935. It was a two-thirds-scale model of the John Deere A, which came out in 1934. The B had many of the same features and appearance of its big brother, such as a four-speed transmission, a hydraulic power lift, adjustable rear wheel tread, rear PTO, and one-piece rear axle housing. The B cost less than the A and used less fuel (kerosene). It was designed to replace a team of horses and was capable of pulling one 16-inch or two 12-inch plow bottoms. Rubber tires were available as an option from the beginning. The two-cylinder transverse, horizontal, side-by-side engine displaced 149 ci.

From the first serial number (1000) to serial number 42199, a shorter frame was used. After that, the long-frame versions were capable of using the same cultivator equipment as the A. The B went on to be the largest-selling model in Deere's history with over 325,000 sold through 1952. Variations, besides short and long frames, include styled and unstyled, general purpose and standard-tread, Hi Crops, wide-front and single-front-wheel general purpose, industrial, and orchard models. The Lindeman Brothers of Yakima, Washington, made a crawler modification that became quite popular.

The rarest is the 1935 B with the original four-bolt front pedestal. After serial number 3042, the four-bolt pedestal was discontinued in favor of an eight-bolt design.

1934 John Deere B. The B was about two-thirds the size of the A and had similar features. The one pictured is the BN version with a single front wheel. Rubber tires were an option. *Andrew Morland*

JOHN DEERE
GENERAL PURPOSE
REG. IN U.S. PAT. OFF.

1936 FARMALL F-30

The Farmall "Regular" had been in production for nearly seven years when International Harvester decided that the time had come for a larger version to be unveiled: the Farmall F-30. At the same time, the Regular became the F-20. The F-30 (basically, 30 horsepower) was built from 1931 to 1939 without much change, except for the switch from gray to red paint. The cable steering brakes were retained from the Regular, and the engine was the same 284-ci four-cylinder unit that was used in the McCormick 10/20 standard-tread tractor but was rated at 1,150 rpm, rather than 1,000 rpm. The F-30 was rated for three 14-inch plow bottoms and was said to be able to plow an acre per hour. Steel wheels were standard, but rubber tires were optional. When rubber tires were specified, a faster fourth gear was provided in the transmission or could be added later. A hydraulic implement lift was added in 1938. The F-30 was longer and heavier than the Regular. It weighed about 6,000 pounds.

Besides being historically significant as the first variation of the basic Farmall, there are several other interesting variations on the original F-30, including wide and narrow front- and rear-tread widths, high-crop, and rice and sugar cane specials.

1936 Farmall F-30. The bigger Farmall came out in 1931, a three-plow version of the previous two-plow Farmall Regular. The rubber tires, on French & Hecht wheels, were installed in 1943 and are still in use. *Ralph W. Sanders*

F-30
McCORMICK-DEERING
FARMALL

1937 OLIVER HART-PARR 70 ROW CROP

The Oliver 70 (1935–48) represents a significant milestone in tractor history, as it was one of the first, if not *the* first truly "styled" tractor. It was also one of the first to be offered with a modern self-starter and lights, and it was touted as being easier to drive than previous machines, as it had steering column controls for its smooth-running six-cylinder 216-ci engine. The "70" designation represented the gasoline octane rating required for the high-compression engine; a kerosene-burning, low-compression engine was also offered. It was available in row crop, standard-tread, and orchard models. In its University of Nebraska Test, No. 252 in April 1936, the gasoline version pulled 89 percent of its own weight. Rubber tires were a popular option. The 70 weighed 3,460 pounds and had a top speed in fourth gear of 6 mph.

About 5,000 of this original version were sold in 1937. In late 1937, it was updated and restyled as part of the Fleetline series. A six-speed transmission replaced the original four-speed unit at that time. At that time, the Hart-Parr name disappeared. All versions were also sold in Canada by the Cockshutt Farm Equipment Company of Brantford, Ontario, in their livery as the Cockshutt 70.

1937 Oliver Hart-Parr 70 Row Crop. One of the most interesting innovations associated with the Oliver 70 of 1935 was the "Tractor Color Voting Contest." Six colors were shown to customers to vote on. The one shown was painted "Regatta Red." Green with red trim was eventually selected. *Ralph W. Sanders*

1937 FORDSON ALL-AROUND

In 1937, the Fordson tractor celebrated its twentieth birthday with a new version: the tricycle-configured Fordson All-Around. Production in Dagenham, England, was split between the All-Around and regular Fordson Ns. The All-Around was an attempt to get in on the move to all-purpose or general-purpose row-crop tractors that were becoming popular in America. It was also an effort to regain the market lost when Fordson production ceased in the United States.

In the 1930s, the concept of the all-purpose tractor came from the farmer's desire to replace not just some of his horses with a tractor, but to replace *all* of his horses with a tractor. A characteristic of all-purpose tractors, following the lead of the International Harvester Farmall, was the tricycle configuration that dispensed with the low front axle. This made front-mounted cultivators practical for taller crops such as corn. Horse-drawn sulky cultivators were prevalent at that time. Another feature of the all-purpose tractor was the rear-center power takeoff (PTO) to drive towed harvesters. Formerly, horse-drawn harvesters were powered by gears driven by one of the machine's wheels, hardly a satisfactory arrangement. The Fordson All-Around had all of these features.

The Fordson All-Around was built in England from 1937 to 1940, but almost all were shipped to the United States in 1937 and 1938. The 1937 models were painted blue with orange wheels and had the Case-style chicken-roost steering. For the rest, the paint scheme was all orange, and a complex series of shafts and universal joints reached a worm and sector gear on the front pedestal to affect steering. All of the All-Arounds are considered rare since their numbers are limited. Their performance numbers are virtually the same as for the Fordson N. Serial numbers for the All-Around were not separated from the regular Ns, so it is not possible to identify them by serial number.

1937 Fordson All-Around. Imported from England to meet the demand for all-purpose tractors, this was Ford's first attempt at a tricycle tractor. There were at least two steering arrangements, but this machine used the chicken-roost type similar to that used by Case. *Ralph W. Sanders*

1937 JOHN DEERE G (LOW RADIATOR)

A hallmark of the year 1937 was a return to optimism by Midwestern farmers who had been suffering the effects of the Great Depression. Progressive farmers began thinking of adding a second, larger tractor to expand acreage tilled. The new, rugged John Deere G filled that bill.

The new 1937 G was a three-plow tractor that had about the same power as the venerable D but was almost 1,000 pounds lighter. Therefore, more of the power was available to the drawbar and not so much was used to move the tractor itself. Archrival International Harvester had introduced a three-plow row crop machine in 1931 called the F-30. Deere initiated studies of a three-plow machine to be designated the F. The "F" designator was carried through several years of gestation, but just before the tractor was released for production, the change was made to G. The stated reason was to avoid customer confusion with IH models, the F-20 and F-30.

Besides three 14-inch plow bottoms, the new G could handle a 28-inch thresher or four-row cultivators. A four-speed transmission was included with a single lever operating in a cast gate, rather than the two-lever arrangement used on A and B tractors. Throughout its life, the G's 412.5-ci engine was meant to burn distillate fuel. Gasoline conversions were offered "aftermarket" which could put the G at over 50 horsepower.

The first Gs entered the field in May 1937. Soon thereafter, complaints began coming in from hot-weather areas that the G was overheating. Deere engineers quickly designed a taller radiator that required a notch in the tank top for the steering shaft to pass through. At Serial Number 4251, tractors left the factory with the new radiator. Deere also offered new radiators to owners of earlier tractors. Most farmers took the larger radiator, whether or not they were experiencing overheating. Today, an original early G with an authentic low radiator is much sought after by collectors of antique Deere tractors.

1937 John Deere G (Low Radiator). Soon after the G entered service in 1937, complaints began coming in from farmers in warmer climates that the tractor was overheating. Deere engineers quickly designed a taller radiator that required a notch in the tank top for the steering shaft to pass through. Those that escaped the recall are now especially collectable and are referred to as low-radiator Gs. *Ralph W. Sanders*

1941 FARMALL MD

It was a big surprise in 1941 when a diesel engine version of the big Farmall M was introduced, called the MD. International Harvester had pioneered diesel wheel tractors when the WD-40 was introduced in 1934. Only Caterpillar had preceded IH when the first diesel tractor, the Caterpillar Diesel Sixty, came out in 1932. By 1941, the diesel engine was well accepted as stationary power plants, in ships, crawlers, and even dirigibles, but they were not yet common in trucks or other wheeled vehicles. Caterpillar introduced its DW-10 diesel wheel tractor, teamed with an earthmover scraper, in 1940, but the Farmall MD was history's first diesel row-crop farm tractor.

The block part of the 248-ci four-cylinder overhead-valve engine was the same as that in the regular M. For the diesel, aluminum pistons and a five, rather than three, main bearing crank were used. The main difference was in the head. Besides having twice the compression ratio of the gasoline version, the head contained the IH starting system. The engine was started as a gas engine, and when warm, it could be switched over to run as a diesel.

A lever at the platform accomplished the switch over. With the lever pulled, the combustion chamber was enlarged to provide a compression ratio of 6.4:1. Intake air came in through a carburetor, gasoline was allowed to flow into the carburetor as well; an ignition circuit was completed, providing fire to spark plugs. After a minute or so on gasoline, actuation of the lever shut all that off and increased the compression to 14.2:1. With the diesel intake open and the injectors energized, the engine now ran on diesel fuel.

Large-acreage farmers found the MD paid for its initial cost over a conventional M in less than a year since it used about a third less fuel at full power, and diesel in those days cost a little over half as much as gasoline. Fuel consumption savings were even greater over gasoline and distillate at part loads.

Other than the engine, the MD was the same as the conventional M, with a five-speed transmission, optional wide-front or tricycle configurations, and the same wheel-tire options.

1941 Farmall MD. The first diesel row-crop tractor, the Farmall MD, originally cost about $450 more than the gasoline version but saved that amount in fuel cost for the average farmer in about a year. The Farmall styling by famous industrial designer Raymond Loewy is still quite fresh-looking today. *Ralph W. Sanders*

1955 MINNEAPOLIS-MOLINE GBD/MASSEY-FERGUSON 95

The Minneapolis-Moline G made its appearance in 1950 with a 403-ci four-cylinder gasoline engine. By 1955, the GBD (Diesel) version was introduced with a 425.5-ci six-cylinder diesel. The engine was of the Lavona-type with a precombustion chamber. At almost 10,000 pounds working weight, it was one of the heaviest wheel tractors ever tested at the University of Nebraska. The M-M GBD was built from 1955 to 1959. It had a five-speed transmission giving it a top speed of 15 mph.

In 1958 the management of Massey-Ferguson elected to buy M-M GBDs and put its grille style and paint on them to offer the higher-horsepower diesel tractors their customers wanted. This version was sold as the Massey-Ferguson 95 through 1960, although after 1959, M-M had changed from the GBD to a Gvi version with the engine speed raised from 1,300 to 1,500 rpm, thus raising the horsepower from 63 to 78. For 1960, Massey 95s were based on the M-M Gvi. Any of these are big, colorful, and impressive milestone tractors indicative of diesel tractor technology in the mid-1950s. In the 1960s, the M-M G was resurrected as the 100-plus-horsepower G-706.

1955 Minneapolis-Moline GBD. The first M-M diesels came out in 1954 featuring the Lanova system of fuel injection and combustion. This design used an energy cell opposite the injector, which was the major principle of Lanova's "Controlled Turbulence" combustion chamber. *Ralph W. Sanders*

MM
DIESEL
MINNEAPOLIS-MOLINE GB

1956 MASSEY-HARRIS 333

In mid-1953, Harry Ferguson was trying to develop a tractor-mounted harvester that could be installed and removed like any other implement. Since Massey-Harris was the industry leader in harvesters, Ferguson sought help from Massey in its design and production. Thus began talks that would eventually lead to an amalgamation of the two industry giants toward the end of 1953. After the merger, the two continued operations for a time as separate companies. Ferguson engineers updated their Detroit-built tractor in 1955, and so did Massey with those built in Racine, Wisconsin.

The Massey-Harris 333 replaced the 33 in 1956 with a new bronze engine color scheme; some chrome trim was added to the grille. The four-cylinder overhead-valve engine displacement was increased from 201 ci to 208 ci, giving it 40 PTO horsepower. It gave a drawbar pull of 77 percent of its ballasted weight of

7,000 pounds. The 333 had built-in gasoline, distillate, diesel, and LPG versions, and in standard-tread or row-crop with either wide-front or tricycle configurations. A live PTO was an option, as was power steering and a draft-control three-point hitch. None of these versions sold in large quantities. Fewer than 3,000 of all types were delivered in 1956 and 1957 and are considered rare—especially the diesel version, with only 500 delivered.

Of historical significance, Massey was the first in the industry to go to 12-volt electrical systems. A 12-volt electrical system was standard on all Massey models including the 333. The main improvement, besides the electrical system on the 333, was the addition of a two-range shift, giving a total of ten speeds forward and two in reverse.

1956 Massey-Harris 333. Massey and Ferguson were still operating as separate companies in 1956, but the 333 got the Ferguson draft-control three-point hitch. The 333 was also one of the first farm tractors to use a 12-volt electrical system, but a generator was still used, rather than an alternator. The tractor featured power-adjustable rear wheels. *Ralph W. Sanders*

1963 ALLIS-CHALMERS D-19

Despite working with outdated manufacturing facilities, Allis-Chalmers produced creditable and competitive products. The Allis-Chalmers D-19 was no exception. It is historically significant because the diesel version was the first production tractor to use a turbocharger, which has since become almost standard fare among diesel tractors. It was also available with LPG and gasoline engines, but only the diesel incorporated the turbocharger. Only about 10,000 of all types were built. All of the engines were six-cylinder types of 262 ci. It was nominally a 65-PTO-horsepower tractor. Interestingly, Allis-Chalmers was one of the largest producers of turbochargers for aircraft in World War II.

The D-19 was only built from 1961 to 1963, with the diesel version being the most popular and the LPG type the least. It was a five- to six-plow tractor with a live PTO and a three-point hitch. Power steering was standard. A fixed-ratio transmission, including a partial-range power-shift, was used on all versions. It provided eight speeds forward and two reverse and a top speed of 14 mph. Weight was about 6,800 pounds without ballast.

1963 Allis-Chalmers D-19. The diesel version of the D-19, which was the most popular fuel option, was the first farm tractor to be equipped with a turbocharger. *Ralph W. Sanders*

A-C
ALLIS-CHALMERS
ALLIS-CHALMERS
D19

Distinctive HORSEPOWER

THE FOLLOWING TRACTORS ARE EXCEPTIONAL AND ATYPICAL among their peers. Most were built in relatively small lots; some were actually prototypes for later production in volume, such as the 1936 Ferguson-Brown, which lead to the famous Ford-Ferguson. Orchard tractors are distinctive because of their flowing, race car–like lines. In some cases, the companies that made these tractors were short-lived, leaving their progeny orphaned. The Ford BNO-40 aircraft tug, on the other hand, was made by the thousands, but most were junked after World War II. Nevertheless, all of the following tractor models have a story and did important work during their lives.

- 1933 John Deere GPWT
- 1935 Kaywood D
- 1936 Ferguson-Brown Type A
- 1936 Silver King R-66
- 1937 Love Tractor
- 1937 John Deere AOS
- 1938 Massey-Harris 101
- 1938 Avery Ro-Trak
- 1939 Ford-Ferguson 9N (Aluminum Hood)
- 1940 McCormick O-6
- 1942 Ford BNO-40 Aircraft Tug

1933 JOHN DEERE GPWT

Tractor competition in the late 1920s was quite robust, with the Fordson number one in sales; International Harvester was second, with John Deere third with its only model, the D. International Harvester, with its general or all-purpose Farmall, was setting a trend toward tractors with mounted cultivators. With its tricycle configuration, the Farmall could cultivate two rows at once, but for plowing, the tricycle configuration was a disadvantage because its right front wheel could not run in the previous furrow and act as a guide. Deere's answer was the GP (for General Purpose), which originally had a wide front with an arched front axle. In theory, it could straddle one row, cultivating three at a time and retaining the wide front, making it better at plowing.

This theory did not gain popularity with the farmer, however, and by late 1929 Deere brought out the GPWT (General Purpose Wide-Tread), a tricycle tractor with 76-inch rear tread. At first, the steering shaft ran along the right side of the engine (serial numbers 400000–404809). Then in 1932, the steering was changed so that the shaft ran over the engine (like the Farmall). Serial numbers for this final version of the GPWT are 404810–405252. Production ended in 1933 for the GPWT but continued through 1935 for the standard-tread GP (which had its own serial number sequence).

The GP was unique among Deere tractors in that its two-cylinder horizontal side-by-side transverse engine was of the L-head, or side-valve, configuration. Originally, it displaced 312 ci, but in 1931, displacement was increased to 339 ci. It had a three-speed transmission, which gave it a top speed in third gear of about 4 mph. Originally, the weight was 4,260 pounds, but versions after 1931 had grown to a little over 4,700 pounds.

1933 John Deere GPWT. This is the tricycle row-crop version of Deere's Model GP wide-front. The original GPWT came out in 1929 with side steering shaft running alongside the engine. For this 1933 version, Deere incorporated over-the-engine steering. *Ralph W. Sanders*

1935 KAYWOOD D

One of several little-known orchard tractors made in Benton Harbor, Michigan, the Kaywood D OX-85 Orchard was produced from 1935 to 1937 with local apple and cherry growers in mind. It was a four-wheel, standard-tread design, related to a very similar tricycle tractor made by Parrett Tractors, also of Benton Harbor, Michigan. Both have the same grille work over the radiators and both use 133-ci Hercules IXB four-cylinder engines of 3.25 x 4-inch bore and stroke. Both have three-speed, shift-on-the-fly, automobile-type transmissions with fairly high road speeds: 17 mph for the Kaywood D and 20 mph for the Parrett. Both had optional starters and lights, and both weighed just under 3,000 pounds.

Dent Parrett, founder of Parrett Tractors, began in the tractor business in Chicago in 1916 with a four-wheel machine that used a four-cylinder Buda engine of 25 horsepower. His next endeavor was a motor-cultivator in 1919. The year 1930 found Parrett in the Benton Harbor fruit-growing area, making tractors and wooden crates for the fruit industry. It was at this point that he came in contact with an engineer named Jabez Love, who went on to make the Love Tractor, which later morphed into the Friday Tractor.

1935 Kaywood D. Made by the Kaywood Corporation for fruit growers, the Model D was a four-wheel, standard-tread design using a Hercules IXB four-cylinder engine of 133 ci displacement. It had a shift-on-the-fly, automobile-type transmission with fairly high road speed of 17 mph. *Ralph W. Sanders*

1936 FERGUSON-BROWN TYPE A

Harry Ferguson was a complex man. His personality, according to historians, embodied subtlety, charm, rudeness, modesty, brashness, and petulance—and the switch between could be abrupt and unpredictable. Nevertheless, the plucky Ulsterman (Northern Ireland-Protestant) cut a fairly wide swath through the agricultural history of the twentieth century. In fact, his name still graces some of the best of twenty-*first*-century farm machinery.

His tractor experience began in the First World War when he was assigned by the government to oversee efforts to increase domestic food production by managing government-owned tractors operating in Ireland. After the war, he, with some talented associates, invented his famous hydraulic three-point hitch. He first applied this to the venerable Fordson and expected Ford's production genius Charles Sorenson to jump at the chance to produce it. When that didn't happen, he made his own tractor to demonstrate the system to other manufacturers. Painted black, this tractor became known as the "Black Tractor." A deal was struck with gear-maker David Brown of England to make a tractor that became known as the Ferguson-Brown Type A.

The production version looked like a scale model of the Fordson, about 80 percent of its size and about half of its weight at around 1,700 pounds. It had a conventional differential (rather than the worm drive of the Fordson), individual brakes, and a four-cylinder engine of 123 ci from the Climax automobile company. It was painted industrial gray.

The Ferguson-Brown Type A was made by David Brown from 1936 to 1938. When considering the cost of the special three-point implements, it cost twice as much as the Fordson and did not sell very well. Brown wanted to start over with a larger tractor; Ferguson thought the problem was insufficient production volume—and who knew most about mass production? Henry Ford. So Ferguson crated up a Type A and implements, and sailed for America to meet with Ford. The result was the highly successful Ford-Ferguson 9N.

1936 Ferguson-Brown Type A. In the late 1920s, Harry Ferguson built a tractor of his own design with his newly invented three-point hitch. Tests of this tractor were impressive enough that the David Brown Company agreed to a production deal. The gray production version used a Climax engine. *Ralph W. Sanders*

1936 SILVER KING R-66

1936 Silver King R-66. The round hole in the front casting is for mounting the cultivator on this all-purpose tractor manufactured by the Fate-Root-Heath Company of Plymouth, Ohio. *Ralph W. Sanders*

Made between 1935 and 1939 by the Fate-Root-Heath Company (an old-line railroad locomotive builder) of Plymouth, Ohio, the Silver King was the company's second tractor model. The first, called the Plymouth, was essentially the same, but the second was painted silver and called the Silver King. It was powered by a Hercules four-cylinder L-head engine of 133 ci, giving the tractor a belt-power rating of 20 horsepower. Some early advertising showcased a Continental engine, but when tested at the University of Nebraska (Test No. 250) in 1935, documentation specified the Hercules engine. A four-speed transmission was used with a speed of 14.5 mph at rated engine speed of 1,400 rpm, but with a governor override, the top speed was a brisk 25 mph, making it the fastest tractor of its time.

The R-66 model number indicated a rear-wheel tread width of 66 inches. This version used a single front wheel and had bull gears on the rear axle to give clearance for row-crop work. Pads for cultivator mounting were provided. The Silver King was rated for two 16-inch plows.

Various models of Silver King tractors were produced in small numbers through 1956. All are considered to be rare and highly collectible.

The Silver King used a unique covered steering linkage and a gooseneck frame. A governor override throttle gave transport speeds of up to 25 mph. *Ralph W. Sanders*

1937 LOVE TRACTOR

Jabez Love was an engineer working for Dent Parrett at a Benton Harbor, Michigan, tractor and fruit-box factory in the early 1930s. He noticed that fruit growers often used horses and wagons to carry their produce out of the orchards, and then transferred it to trucks to haul it to the marketplace in Benton Harbor. The orchard trees were too close together for the trucks, or indeed tractors, except for specialty orchard machines such as those made by Parrett, to go between. With Parrett's help (he had a history of making tractors in Chicago in the 1920s), Love made what he called the "Trucktor" from B Ford components. It was narrow and streamlined enough to go through the orchards but had a small truck bed to carry fruit all the way to the market at speeds of up to 40 mph. Love manufactured the Trucktor from 1933 to 1937.

In 1937 Love modified the Trucktor into the Love Tractor, realizing that the narrow, rubber-tired wagons could carry more fruit per trip than his Trucktor bed. He continued with the B engine, Ford truck transmission, and the modified rear-axle assembly. A Fordson front axle and steering assembly were also used. The transmission was a Ford four-speed unit with a two-speed auxiliary mounted behind the transmission. The "bull-nose" grille and sheet metal were added. Holes were cut in the sides of the hood so that headlights could be folded in for protection.

The 1937 Love tractor pictured, serial number 37 O 121, is owned and was restored by Rich Buzalski;

37 indicates the year, O is for Orchard, and 121 is the production sequence. It is unknown if the sequence began at number 1, whether it was for orchard tractors only, or if it started with the Trucktor. The previous owner, a fruit grower, had replaced the four-cylinder B Ford engine with a post–World War II Ford truck flathead V-8, and it will indeed reach speeds of more than 60 mph, but handling qualities and brakes make 40 mph the practical upper limit.

In 1939 Love modified the design using Dodge truck components; then, most models had a five-speed transmission and a two-speed rear axle. Love also became a dealer for Ford-Ferguson tractors. The hydraulic three-point hitch was such an attractive feature that Love copied it (despite the patents) and helped Sears apply it to the Graham-Bradley tractor (without the draft control feature). Love also advised Willys and Dodge in adapting the three-point hitch to the Farm Jeep and the Dodge Power Wagon.

After World War II, David Friday began making the Friday tractor, which is virtually the same as the 1939 Love. Records indicate that Love built tractors with Chrysler, Ford, and Willys components until 1953.

1937 Love Tractor. This is another of several similar orchard special tractors designed for Michigan fruit growers. These machines were made to slip between the low-hanging tree branches and also to pull wagonloads of fruit to the marketplace. This particular version is powered by a 239-ci Ford Flat Head V-8 and could reach 60 mph. *Rich Buzalski*

1937 JOHN DEERE AOS

Between 1937 and 1940, John Deere made about 800 AOS streamlined orchard tractors with serial numbers from AO-1000 to AO-1801, a different number sequence from those plain AO tractors without the stylish sheet metal that allowed the AOS to slip through orchard foliage. These were based on the standard-tread A tractors but had low seats and differential brakes and did not have protruding exhaust pipes or air intakes.

Besides being few in number, the AOS model was heavier and lacked the hydraulics of the row-crop A. Otherwise, the same 309-ci, two-cylinder side-by-side transverse engine and four-speed transmission were used. An interesting and unique feature of the AOS was a special clutch lever that was pulled up rather than forward to engage, because of its cramped cockpit. Electric starting was an option.

1937 John Deere AOS. The Model A, Orchard, Streamlined, had a low seat and steering wheel for the operator. Also, protrusions, such as the radiator cap, were smoothed over, so the tractor could slip through the limbs and branches. *Ralph W. Sanders*

1938 MASSEY-HARRIS 101

The Canadian Massey-Harris Company was enduring hard times during the Great Depression of the early 1930s. It needed new tractors, but some in management favored retrenching and even eliminating the Racine, Wisconsin, tractor operations. Fortunately, cooler heads prevailed and inaugurated a new era in Massey-Harris history with the creation of the great Chrysler-engined tractor series.

Most tractors of the times were either two- or four-cylinder machines, except for Oliver, which was having success with its smooth-running, six-cylinder 70. As time and money were not available for Massey-Harris, the availability of a Chrysler power plant was just what the company needed. Further, Chrysler engine parts and service were available worldwide.

Massey-Harris introduced the 101 in 1938. Its most striking styling feature was undoubtedly the louvered side panels; these proved to be impractical, and the company soon discarded them. It is, however, the side panels on the tractor that now make these tractors such unique collectibles.

Under the skin, the 101 had an L-head six-cylinder Chrysler industrial engine of 201 ci. Some said it purred with mellow resoluteness. Since electric starting had long been standard, these engines came so equipped; thus the 101 was the first tractor to offer a starter as standard equipment.

Twin Power was a Massey-Harris feature employed on the 101. Drawbar engine speed was governed at 1,500 rpm, while engine speed for belt work was 1,800 rpm. This minimized wheel slippage while towing. In top (fourth) gear, 1,800 rpm was available for road travel at up to 18 mph.

1938 Massey-Harris 101. A stylish grille and louvered side panels were the hallmark of the Massey-Harris 101. *Ralph W. Sanders*

MASSEY-
HARRIS
Goodrich
Silvertown

There were two Avery tractor companies: B. F. Avery & Sons in Louisville, Kentucky, and the Avery Company in Peoria, Illinois. Both had a long history in the agricultural equipment industry. The Peoria company, however, produced a unique tractor from 1938 to 1941: the Avery Ro-Trak.

The Ro-Trak was readily convertible from wide-front to tricycle configuration, without the necessity of readjusting wheel alignment. You could simply lift the front, pulling pins and swinging the front wheel castings in or out, for tread widths of 10 inches to 60 inches. The vertical castings contained coil springs that allowed the front wheels to go over bumps without twisting the frame. The Ro-Trak was not tested at the University of Nebraska but was rated for up to three 14-inch plow bottoms.

The company used a Hercules six-cylinder L-head engine of 212 ci. A three-speed transmission gave 16 mph in top gear. Electric starting was standard, but lights and a belt pulley were optional. The Ro-Trak weighed 4,000 pounds.

It is not known how many Ro-Traks were made, but enough, apparently, to keep the company afloat until the start of World War II, when it folded.

1938 Avery Ro-Trak. The feature of the Ro-Trak was that it was readily convertible from wide-tread to tricycle configurations. Otherwise, it was a conventional tractor. *Ralph W. Sanders*

The Ro-Trak was a two-plow tractor driven by a Hercules six-cylinder engine. The unique convertible front end feature never caught on with farmers. *Ralph W. Sanders*

1939 FORD-FERGUSON 9N (ALUMINUM HOOD)

The Ford-Ferguson tractor was birthed in 1939 following a handshake agreement between Irishmen Henry Ford and Harry Ferguson, with Ford providing the production genius and Ferguson providing his patented three-point hitch with draft-control. The result was what has become known as "the tractor of the twentieth century." The 9N tractor, itself, is not so collectible because there were so many made (more than 10,000 in 1939 alone). But the early '39s, with aluminum hoods, grilles, dashboards, and some casting covers, are particularly collectable and rare since only the first 700 or so were made that way. Further, because these items were so fragile and easily broken, many have not survived and were replaced with steel. Aluminum castings were used in the first place because steel stamping equipment was not available in time for the first production batch.

The little Ford-Ferguson is referred to as the tractor of the twentieth century because it started the trend away from row-crop tractors and toward what was to become known as the "utility-configuration." Secondly, the draft-control, three-point hydraulic hitch would eventually become the standard of the industry. Ford sold the tractors to Ferguson, who had the dealerships and marketed the tractors and implements. Ford and Ferguson's partnership split after Henry Ford died in 1948 and his grandson, Henry Ford II, had taken over the company. He abrogated the handshake agreement after finding that Ferguson was making a profit while Ford was losing money on every tractor. An ensuing lawsuit over Ford's use of Ferguson's patents in the 8N tractor built without Ferguson from 1948 to 1952 was eventually settled out of court, but the 1953 Ford tractor had a different hydraulic system, although it still had the three-point hitch—the patent on that having run out by then.

The engine of the 9N was a four-cylinder L-head of 119.7 ci. A three-speed transmission was provided, but a step-up auxiliary was available in the aftermarket, giving a top speed of about 18 mph. The tractor weighed 2,500 pounds and was rated at 12 acres per hour plowing with two 14-inch mounted plows.

1939 Ford-Ferguson 9N (Aluminum Hood). The early '39s were made with aluminum hoods, grilles, dashboards, and casting covers, because steel stamping equipment was not available in time. *Ralph W. Sanders*

1940 McCORMICK O-6

The first variations of the striking Farmall series introduced in August 1939 were the standard-tread running mates of the Farmall models H and M. McCormick labeled these W-4, the equivalent of the H; W-6, the equivalent of the Farmall M; and WD-6, the diesel version. Further variations on the theme were orchard (O) models, industrials (I), specialty models (S), and some combinations. Then in 1952, a Super W-6 and its variations came out with a 264-ci four-cylinder engine in place of the original 248-ci unit. These could have the Torque-Amplifier (TA) two-speed power-shift added to the regular five-speed transmission. Also, in Great Britain, the Doncaster factory turned out a very similar line with the prefix B in the model letters, for Britain.

The tractor pictured is typical of the original 1940 O-6 orchard version of the W-6. These tractors featured sweeping and pointed fenders covering the rear wheels and a tapered shield to offer the operator a measure of protection from overhanging limbs. Air cleaner intakes were also shielded, and the exhaust came out underneath the machine. The headlights were under the grille. Rubber tires were standard equipment. Electric starting was an option on the spark-ignition versions. The O-6 weighed about 4,700 pounds. An OS-6 version did not have the rear-wheel coverings.

1940 McCormick O-6. The orchard version of the Farmall Model M, the O-6 had orchard shielding that let it slip through the trees without damaging blooms, trees, or fruit. The driver sat in a low seat with a protective shield. *Ralph W. Sanders*

1942 FORD BNO-40 AIRCRAFT TUG

Ford based the BNO Aircraft Tugs on the Ford-Ferguson 2N. There were two styles: the BNO-25 (single rear wheels, rear axle, and differential from the 1-ton pickup) and the BNO-40 (dual rear wheels, axle, and differential from the 1½-ton truck). The model designation indicated the weight of the airplane they were designed to pull; 25,000 pounds for the BNO-25 and 40,000 pounds for the BNO-40.

Ford built the BNOs low to the ground; they had a single brake pedal and hydraulic brakes on the rear wheels. There was no three-point hitch or PTO. Heavy fenders almost doubled the weight of the BNO-25 and quadrupled it for the BNO-40.

The same L-head four-cylinder engine used in the 9N and 2N Ford-Ferguson tractors powered the Tug. Displacement was 119.7 ci (exactly one-half of the 239-ci V-8). The tractor used the same three-speed transmission. Weight estimates for the BNO-25 and BNO-40 are 4,050 pounds and 10,000 pounds respectively.

While there were many of both styles made, not many have survived.

1942 Ford BNO-40 Aircraft Tug. Made for the military in World War II, and based on the 2N Ford-Ferguson, the BNO-40 could maneuver heavy aircraft around on the flight line. The BNO-40 was built low to the ground so it could slip under the wings of aircraft on the ramp. It used components of the Ford-Ferguson 2N and Ford trucks. A lighter model, the BNO-25, had single rear wheels. *Andrew Morland*

IMPORTANT !

The LITTLE Guys

THERE WAS A *NEW YORKER* MAGAZINE CARTOON some years ago showing a guy wearing a straw hat driving a riding lawn mower. Above his head was a "thought balloon" that pictured him driving a huge John Deere. But not all tractor owners dream big—small tractors have their charms. Everything about them costs less than for the big boys. Tire prices, for example, go up exponentially with size. Then there is storage space, hauling problems, and special-equipment requirements for lifting parts, such as an 800-pound rear wheel and tire. Standard eight-foot-high garage doors are sometimes not adequate for bigger tractors with vertical exhaust stacks. And if you want to work your oversize tractor, you need big jobs for it to do, as well as big implements. Big is nice for shows, parades, and tractor rides, but there are downsides.

The following machines showcase collectible tractors on a smaller scale.

1932 CATERPILLAR TEN

The Cat Ten is the smallest Caterpillar ever built, and there were only about 5,000 of them made. The Ten weighed in at less than 5,000 pounds, which makes it easily hauled behind a ¾-ton pickup truck. There were high-clearance and wide-track models, as well as versions with electrical systems and rear belt pulleys. Unique among Caterpillars, along with its sibling, the Cat Fifteen, is the use of an L-head engine; all others have overhead valves. Fuel for the Ten was gasoline.

The Cat Ten was about the same size as its predecessor, the Holt T-35 cum Caterpillar 2 Ton but was otherwise completely different. Its engine was smaller: 143 ci versus 251 for the 2 Ton. Also, the Ten used dry clutches while those of the 2 Ton were wet. Both the Ten and the 2 Ton were rated for two 12-inch plow bottoms.

Correct paint can be either gray with red letters or yellow with black letters.

1932 Caterpillar Ten. The Cat Ten was made from 1928 to 1933 and was the smallest Caterpillar made. It had a tread with of 37.5 inches. The one shown has rubber track pads installed. *Ralph W. Sanders*

1934 McCORMICK-DEERING W-12

The W-12 is a standard-tread version of the row-crop Farmall F-12. It is essentially the same, except for conversion to the lower, fixed-tread configuration. It was popular where crop cultivation was not a requirement. The W-12 was not only available as a standard small plowing tractor, but it could be ordered as an O-12 Orchard tractor with sweeping fenders, a Fairway-12 for golf courses and airports, or an I-12 Industrial with hard rubber tires. All versions are considered rare since there were only about 4,000 of the standard W-12 made, plus 4,000 of the O-12s, 600 Fairway-12s, and 1,600 I-12s.

An overhead-valve four-cylinder engine of 113 cid powered the W-12. It had a three-speed transmission, allowing for a top speed of just under 4 mph. The W-12 was rated for one 16-inch plow or two 12-inch plows. Production ran from 1934 to 1938. The standard tractor on steel wheels weighed about 3,400 pounds. Fuel could be either kerosene or gasoline. Rubber tires were available as an option on later-year models.

1934 McCormick-Deering W-12. Using the same engine as the row-crop Farmall F-12, the W-12 could also be ordered for either gasoline fuel or kerosene. The engine had a bore of 3 inches and a stroke of 4 inches. *Ralph W. Sanders*

As a plowing tractor, the W-12 was rated for one 16-inch bottom or two 12-inch bottoms. It was noted for its maneuverability and was also available with rubber tires. *Ralph W. Sanders*

1936 CASE RC

In 1935, International Harvester announced its smaller general-purpose tractor, the F-12. Tractor makers were realizing that the under-100-acre farmer could not afford, and did not need, the larger and more expensive models. Not to be outdone, Case jumped in in 1936 with its diminutive RC.

The birth of the RC was not an easy one. By 1934 Harvester dealers were eating Case salespeoples' lunch with their F-12. Leon Clausen, general manager of Case, resisted efforts by Case marketing and branch people who were urging rapid development of a counterpunch small tractor. His point was that such a tractor would just take sales from the more profitable Case CC. He finally acceded on the condition that the new small tractor would be, in his words, an orphan, so as not to cut into sale of larger tractors. The new RC would be painted a lighter shade of gray, and there would be no advertising, only a brochure. He continued to drag his feet, refusing to develop an engine for the RC, instead buying one from Waukesha. He called the RC a "half-tractor" and threatened "weak-kneed salesmen if they sold the RC for jobs 'over its head.'"

The row-crop, or general purpose, Case RC was a one- to two-plow tractor suited to the needs of smaller farms. It was in Case's inventory from 1935 to 1940, after which it was replaced by the SC. The four-cylinder Waukesha L-head engine displaced 133 ci. It had a three-speed transmission at first, giving a top speed of 5 mph. Early RCs, which weighed about 3,400 pounds, had over-the-engine steering, but in 1937, because of complaints, it was changed to Case's trademark "chicken-roost" steering. In 1939, the RC was restyled with new sheet metal, a cast "Sun Burst" grille, and the new Flambeau Red paint. It also got a four-speed transmission at that time. A standard-tread R was also offered.

In the six years of production, only about 17,000 Rs and RCs were built, making them fairly rare, especially the early overhead-steering types.

1936 Case RC. Although the science of ergonomics was in its infancy in 1936, the RC operating controls were all within easy reach of the driver. Rear-wheel tread was adjustable in width from 44 inches to 80 inches. *Ralph W. Sanders*

The RC sold for about $950 in 1936. Like all Case tractors of the time, a hand clutch was featured along with a thermosyphon engine cooling system. *Ralph W. Sanders*

Shown here on 48-inch steel wheels and optional fenders; rubber tires were optional, but the PTO was standard. The Case RC was aimed at the needs of the smaller farmer. *Ralph W. Sanders*

1938 ALLIS-CHALMERS B

The company named after Edwin P. Allis and William J. Chalmers was founded in 1901. Allis was the largest builder of industrial steam engines in the United States. Chalmers was president of Fraser and Chalmers, manufacturers of mining equipment. Allis had died a few years before the merger of the two firms into the Allis-Chalmers Company, but he was well represented by his two sons, Will and Charles, and by his nephew Edwin Reynolds, who held the title of chief engineer in the new company. Headquarters were in Milwaukee, Wisconsin.

In response to the success of the small Farmalls, John Deeres, and the Case RC, Allis-Chalmers decided to enter the small-tractor field with its own version, the B, in 1938. The B was a one-plow machine weighing less than a ton and featuring a wide front with a high-arched axle. It was built on

a torque-tube frame, and every effort was made to keep costs down. The initial selling price was just under $500. Later versions, with rubber tires and an electrical system, were around $600.

It was designed as a replacement for a team of horses, which still provided most of the motive power for pre–World War II farms. The Allis B was the first GP tractor to offer a bench seat. A foot clutch was used with hand brakes on each fender. The 125-ci four-cylinder engine operated on gasoline. A four-speed transmission gave a top speed of 11 mph.

It is interesting to note that it was a $600 Allis-Chalmers B that Henry Ford used for comparison to the Ferguson-Brown tractor on his Fair Lane Estate in 1938. Besides comparing the Allis tractor's performance, Ford came to understand that $600 was his target price.

1938 Allis-Chalmers B. A torque-tube frame enhanced cultivator visibility, and the high-arched front axle allowed straddling the row being cultivated. The wide cushion seat allowed the driver to slide left or right for a better view. Fenders were standard, but starting was by hand crank. *Ralph W. Sanders*

1939 CLETRAC GENERAL GG

Cleveland Tractor Company (Cletrac) brought out its only wheeled tractor, a small, lightweight unit called the General GG. It was built by B. F. Avery & Co. of Louisville, Kentucky, for Cletrac and produced in small quantities until 1946, when B. F. Avery bought the rights and renamed it the Avery A. The A, like the General GG, had only a single front wheel, but in 1946, Avery added a wide-front version called the V. In 1951 Minneapolis-Moline took over Avery and kept producing the little tractors as the M-M BF.

Other than the front-end configuration, little was changed on the tractor over the years. It was built on a conventional rail frame. Like the Allis-Chalmers B, it had a driveshaft in a tunnel to the three-speed transmission with the shift lever between the driver's knees and individual brake levers on the fenders. The steering shaft was slightly angled to the left to pass the engine. A Hercules four-cylinder L-head engine, with 132.7 cid was used throughout. It produced about 20 horsepower. A self-starter and lights were optional. The tractor weighed 2,800 pounds in working trim.

1939 Cletrac General GG. They say success has many fathers, so the little GG must be considered a successful tractor as it carried several brand names besides Cletrac, including mail-order house Montgomery Ward. A crawler version, the HG, was also offered by Cletrac in 1939. The base price for the GG in 1939 was $595 FOB Cleveland. *Ralph W. Sanders*

The General
BUILT BY THE CLEVELAND TRACTOR CO.

1948 ALLIS-CHALMERS G

Known as the "hoe-on-wheels," the Allis G was just what the nurseryman dreamed of. For delicate crop cultivation, it had no peer. Besides the rear-engine configuration opening up unobstructed crop visibility, foot pedals shifted the hoes for minor corrections. Even the steering wheel was segmented like an aircraft control wheel, so as not to block the view. Unlike aircraft, the G's steering wheel was positioned 90 degrees to the right for straight ahead.

The 10-horsepower four-cylinder L-head Continental AN-62 engine displaced 62 ci. This is the same engine that was used in the Massey-Harris Pony. A three-speed transmission provided a top speed of 7 mph. A "special-low" gave 1.6 mph at rated engine speed. An electrical system and hydraulics were options. The engine starter was activated by a pull-cable on the right side of the engine that directly pushed the switch on the starter. One-bottom 10- or 12-inch plows were available. Only about 30,000 Gs were made in the production run from 1948 to 1955.

1948 Allis Chalmers G. This nifty little tractor is equipped with a 12-inch one-bottom plow, a hydraulic lift, a self-starter, and a headlight. *Ralph W. Sanders*

1948 MASSEY-HARRIS 11 PONY

Introduced in 1948, the Pony was one of several competing ultra-small tractors of the time. They were popular with truck gardeners and around estates, golf courses, and large farms. Counting all of its variations, the Pony was the most popular of all the Massey-Harris tractors. The Pony was built between 1947 and 1954. It weighed less than 2,000 pounds and cost less than $1,000. It used a tiny four-cylinder Continental L-head engine of 62 ci, the same engine that was used in the Allis-Chalmers G. The Pony was capable of 11 horsepower on the belt and 10 on the drawbar. A three-speed transmission provided a top speed of 7 mph. Electric starting was standard, but hydraulics was an option. About 29,000 model 11 Ponys were built in North America plus more in France. French versions had more power and even diesel engines.

A 14 Pony came out in 1951 that was the same as the 11, except for the inclusion of a fluid coupling ahead of the clutch. Fewer than 100 of the 14 were built.

A 16 Pacer came out in 1953. It was a Pony on steroids. It looked like the Pony but was 6 inches longer and weighed 400 pounds more (it also cost $400 more). The Pacer had a 91-ci Continental four-cylinder engine, which gave a maximum belt horsepower of 18. The three-speed transmission of the Pony was continued. Approximately 2,800 Pacers were built between 1953 and 1955.

1948 Massey-Harris 11 Pony. The diminutive Massey Pony was not only sold to truck gardeners and small farmers, but it also found use on large operations as a chore tractor. The one shown, despite the snow, is equipped with a sickle-bar mower. *Ralph W. Sanders*

1949 FARMALL CUB

The name "Cub," a departure from the naming norms at IH, had been used quite successfully by Piper and Taylor for their small utilitarian airplanes. It had a friendly ring to it, and it connoted small, cute, and tough. Assembly of the Farmall Cub began in 1947 in the Louisville, Kentucky, plant. Target production was set at 50,000 units per year, although this number was never reached.

The Cub weighed in at about 1,500 pounds without additional ballast and 2,700 pounds as tested at the University of Nebraska. Its initial selling price was $600. The price was not as much of a problem for IH as originally anticipated, since the price of the new Ford 8N had risen to more than $1,000 following the Ford company's split with Harry Ferguson.

The configuration of the Cub was virtually identical to the Farmall A, but was approximately an 80 percent scale model of the A. Unless seen together, they are difficult to tell apart. The easiest way to tell is by the shape of the fuel tank: the Cub's is rounded off, while the A's is more teardrop shaped.

Gasoline was the standard fuel, but a distillate configuration was an option, as was electric starting. The four-cylinder L-head engine (unique to Farmalls) displaced 59.8 ci and produced 10 horsepower. The Cub was equipped with a three-speed transmission throughout its life, which gave it a top speed of 6 mph.

The Cub's life as a Farmall ended in 1958; however, some Cub Lo-Boys were labeled "Farmall" into 1964. Most Cub Lo-Boys were Internationals, as were regular Cubs after 1958. Production of these International Cubs continued through 1975.

1949 Farmall Cub. The quintessential small chore tractor, the Cub set a record, of sorts, for the longest production run of any tractor, continuing from 1947 to 1975 with only minor changes. *Ralph W. Sanders*

1953 MASSEY-HARRIS 21 COLT

1953 Massey-Harris 21 Colt. In a break from tradition, the Massey-Harris Colt had an underneath, rather than vertical, exhaust. The Colt was a 25-horsepower tractor. *Ralph W. Sanders*

The M-H 21 Colt (1952–53) was the old 20 (1946–48) revived with updated styling. The 20 itself was new in number only, for it was identical to the 81 (1941–46). If there was a difference, it was only in the price, which jumped from $500 to $1,200. The number 81 was changed to 20 to line up with a new numbering system inaugurated at the company's 100th anniversary.

For the 21 Colt, the same 124-ci L-head Continental engine was installed in the same basic chassis but with updated sheet metal. A four-speed transmission was retained offering a top speed of 16 mph. Electric starting was standard, but hydraulics was optional. Distillate fuel and standard-tread versions of the Colt were not offered. Row-crop and utility front ends were available. The Colt weighed about 2,600 pounds.

With its utility front end and if equipped with the optional hydraulics and three-point hitch, the 21 Colt provided good competition for the Ford 8N. *Ralph W. Sanders*

1954 MASSEY-HARRIS 23 MUSTANG

The Massey-Harris 23 Mustang was updated styling on the same chassis as the previous Massey-Harris 22 (which did not have a name). The same 140-ci L-head four-cylinder Continental engine was used. The Mustang K offered the option of burning distillate fuel. Dual-tricycle, single front wheel, and utility front ends were offered. The 23 Mustang was built between 1952 and 1956. Hydraulics and electric starting were standard, but a three-point lift was optional. A four-speed transmission gave a top speed of 13 mph. Weight was about 3,000 pounds.

1954 Massey-Harris 23 Mustang. The Mustang was much like the previous Colt but had the engine bore increased from 3 inches to 3.19 inches, raising the displacement from 124 ci to 140 ci. The Mustang could also be ordered configured for distillate fuel and with optional narrow front ends. *Ralph W. Sanders*

1955 OLIVER SUPER 66 (HALF-SCALE)

This half-scale version of the nifty Oliver Super 66 is the product of the Richardson Brothers' garage in Marengo, Illinois. The brothers, Ray and Ron, build these half-scale tractors as a hobby and also sell them to people who take them to shows and fairs. The Richardsons keep some of their work for custom gardening.

The real Oliver Super 66 was produced from 1954 to 1958. It used the same 144-ci four-cylinder engine as the Oliver Super 55 along with the six-speed transmission. This half-scale version uses a three-speed Farmall Cub transmission and differential and a 12-horsepower Kohler engine. Measurements for their half-scale version were copied from a 1/20 scale toy tractor, which was a very accurate rendition of the full-size Oliver.

The frame is designed to accommodate two different half-scale cultivators, which the brothers also made.

The challenge in a half-scale version, according to Ray Richardson, is getting all the components necessary to fit under the half-scale sheet metal. Their tractor features a full-sized seat, along with a full-sized clutch and brake pedals, smaller steering wheels are used.

1955 Oliver Super 66 (Half-Scale). Radiators for most scale model tractors are mock-ups only when air-cooled engines are used, like in this case. Full-scale generators must sometimes be used even though they appear out of proportion to the finished tractor. *Robert N. Pripps*

1955 ALLIS-CHALMERS WD-45 (HALF-SCALE)

Allis-Chalmers WD-45 (Half-Scale). The Richardson Brothers of Marengo, Illinois, also build half-scale implements, such as two-bottom plows, drags, disks, and cultivators for their half-scale tractors. Some of their creations have hydraulic implement lifts. *Ralph W. Sanders*

The "real" WD-45 Allis-Chalmers came out in 1953 and was produced through 1957, configured for LP, gasoline, or distillate fuels. The diesel version came out in 1955. This half-scale model is the product of Ron and Ray Richardson's "Life-of-Riley Garage" in Marengo, Illinois. The two brothers have been making one or two scale models every year for several years: mostly they are sold to folks wanting to display them at shows, but the brothers have kept some for their own yard work and for custom garden plowing.

They started this project with photos of the actual tractor and a small-scale toy. They also got a parts manual for the real tractor. For the half-scale WD-45, an 18-horsepower Kohler engine was used along with a Farmall Cub transmission and differential. Actual Allis-Chalmers sheet metal was sectioned and re-welded to scale. Brother Ray drove this machine on a 35-mile Tractor Trek. It tops out at 8 mph.

Not So Many MADE

TRACTORS THAT HAD LOW PRODUCTION VOLUME are unique and rare because not many were made, and, in some cases, not many survived. Start-up companies flourished right after World War II but quickly died when the majors got up to speed with new offerings. These are now considered orphan tractors. Others had performance or reliability problems and were soon retired to the fencerows, or scrapped out, leaving only a few to be restored.

1939 CLETRAC E-62

Cleveland Tractor Company, or Cletrac, traced its history back to 1911, stemming from a heritage of steam automobiles and sewing machines. The company began making crawler tractors in 1916, primarily for the agriculture and forestry industries. Unlike Caterpillar crawlers, which declutched and braked a track for turning and did not have a differential between the tracks, Cletrac used a differential and simply braked one track to make a turn. Both approaches had their advocates.

The Cletrac E was made for the row-crop farmer between 1935 and 1944, when its designation was changed to DG. The E was sold in five tread widths from 31 to 76 inches. The E-62 had a tread width of 62 inches. It could be equipped with a PTO and with provisions for a front-mounted four-row cultivator. It was capable of pulling up to four plows.

The Cletrac E used a 28-horsepower Hercules four-cylinder gasoline engine of 226 ci. A three-speed transmission gave a top speed of 4 mph. Weight was 6,100 pounds. A starter and lights were optional.

Oliver Corporation acquired Cleveland Tractor Company in 1944.

1939 Cletrac E-62. The E was the smallest crawler in the Cletrac lineup, which consisted of four diesel and eleven gasoline-powered models. The "62" indicated one of the several the optional track widths in inches.

1939 JOHN DEERE BNH

The BNH was the high-clearance B with a single front wheel. The 1939 Model received the styled sheet metal.

One of the most respected names in the new industrial design field was Henry Dreyfuss of New York. He and his firm had been in improving the looks of products from telephones to kitchen appliances. Deere and Company sought his consultation for restyling their entire tractor line in 1937. Dreyfuss came to Waterloo in the fall of 1937. Within a month he had made a wooden mockup of a stylized B. It was more than just a radiator grille. The styling was indeed functional as now the grille protected the radiator from field debris. The striking hood was slimmer, enhancing visibility of under-mounted cultivators. By late 1938 (for the 1939 model year), newly styled A and B tractors were ready for sale. And sell they did, the B even more so than the A.

Of course, there were some technical improvements as well. First, a displacement increase to 175 ci from 149 ci for the two-cylinder engine was instrumental in increasing torque and therefore pulling power. Compression was also increased, although distillate was still the fuel of choice. Options included the hydraulic lift, normal- and slow-speed four-speed transmissions, and 96-inch wheel-tread rear axles on the BN and BW versions. Shipping weight was 3,500 pounds.

1939 John Deere BNH. The Narrow-High version of the row-crop John Deere B featured large rear wheels and a larger single front wheel to raise it above taller crops. *Ralph W. Sanders*

1948 LONG A

1948 Long A. A very conventional row-crop tractor for the time, the Long A was successfully tested at the University of Nebraska Tractor Test Lab in 1949 (Test No. 410). One of its selling features was that it was equipped with Timken bearings throughout. *Ralph W. Sanders*

Following World War II, due to wartime production restrictions, there was a shortage of available tractors. Traditional tractor manufacturers, slow to convert from war-related production, were unable to fulfill the postwar demand. At that time, there were a number of upstart tractor companies entering the market. One of those attempting to supply the postwar needs of farmers was Long Manufacturing Company of Tarboro, North Carolina. It came out with a traditional row-crop tricycle tractor with adjustable rear-wheel tread widths. It looked quite a bit like the Farmall H. The Long A used a four-cylinder Continental L-head engine of 162 cid configured for gasoline fuel. It produced 32 horsepower at 1,800 rpm. This resulted in a three-plow rating with 14-inch plow bottoms. A four-speed transmission gave a top speed of 13 mph. A starter and lights were options. Weight was 3,250 pounds.

The A was only built in small numbers in 1948. For years after that, Long imported various brands of foreign tractors, some of which were badged with "Long."

1949 SHEPPARD DIESEL SD-2

The R. H. Sheppard Company was founded in a factory purchased in 1937 by Robert Sheppard in Hanover, Pennsylvania. Included in the purchase were the rights to produce existing product lines of the Kintzing wire cloth loom, floor polisher, and gas generator. These products provided the basis of the manufacturing facility and a place where tractors with diesel engines could be developed. As various other uses for the engines were found, the business prospered and the early lines were phased out. In 1940, the company bought a factory on Philadelphia Street in Hanover, adding a foundry in 1943. The company still operates at that location, where it produces power steering equipment for the transportation industry.

Sheppard tractors were built in three sizes: SD-1, SD-2, and SD-3. The size numbers indicated the number of plows for which they were rated and the number of cylinders in the engine. All were powered by diesel engines of Sheppard's manufacture. The Sheppard company stressed simplicity and economy, claiming their diesels burned half as much fuel as comparable gasoline-powered tractors. Weight of the SD-2 was about 4,000 pounds. Eight transmission speeds were provided. Top speed was 12 mph.

1949 Sheppard Diesel SD-2. This tractor was part of one of the earliest wheel-type diesel tractor lines, especially in this small two-cylinder size. The SD-2 engine displaced 142 ci and was rated at 25 horsepower.

The SD-2 had a roller chain final drive, a four-speed transmission (eight with optional auxiliary) and a basic weight of 4,000 pounds. It also had a very curious steering shaft routing that traveled down to the frame, along the frame to the front, then into the middle of the tractor under the engine to a ball-screw steering mechanism. *Ralph W. Sanders*

1950 WARDS WIDE-FRONT

Montgomery Ward, not wanting to be outdone by rival Sears Roebuck and its Graham-Bradley tractor, sold tractors made by Custom Manufacturing Corporation of Shelbyville, Indiana, badged "Wards." This wide-front version was the same as the Custom C. Power was from a six-cylinder Chrysler L-head engine of 217.7 ci. A four-speed transmission was provided and hydraulic brakes were standard. The tractor was rated for three 14-inch plow bottoms and weighed 3,450 pounds.

Custom Manufacturing Company had a colorful history of tractor making dating from before World War II. It surfaced in Shelbyville in 1947, making the Custom C tractor. Production continued to 1949. Beginning in 1950, production was taken over by the Harry A. Lowther Company, also of Shelbyville. In 1953, manufacturing was moved to Hustisford, Wisconsin, and was done by Custom Tractor Manufacturing Company Inc. Now labeled "Custom 96" and "Custom 98," these tractors had 230- and 250-ci Chrysler engines respectively, five-speed transmissions, and Gyrol fluid drives, and were serious contenders in tractor-pulling contests.

1950 Wards Wide-Front. The Wards brand name was a common contraction for the Montgomery Ward mail-order company. *Ralph W. Sanders*

The tipping hood gave complete access to the Chrysler industrial six and accessories. This type of hood was used by Harry Ferguson on his TE/TO-20 tractors of 1948. *Ralph W. Sanders*

With its smooth-running Chrysler six engine, the engine's exhaust music could be enhanced with a split manifold and twin straight pipes. *Ralph W. Sanders*

1953 JOHN DEERE 70 LP

The 70 made its debut a little later than its stablemates, the 50 and 60, and it shared their features and styling. It was the successor to the great G but was available in both standard-tread and row-crop configurations. It was also available in a variety of fuel choices besides the All-Fuel (distillate) option of the G. The 70's power beat that of the G's by almost 20 percent, even more for the fuel options other than distillate.

New to the 70 were live hydraulics and a live PTO with a separate clutch, factory power steering, an adjustable seat backrest, longer clutch and throttle levers, and a new rack-and-pinion method of changing wheel spacing. A 12-volt electrical system was standard on all but the diesel version (which used a pony motor starter).

The 70 was only made from 1953 to 1956, so total production of all types was a meager 44,000-some units. Of these, the LP (Liquefied Petroleum) version is probably the rarest. Its two-cylinder engine displaced 379.5 ci. The six-speed transmission gave a top speed of 13 mph. Weight for the dual-tricycle LP version was 6,215 pounds.

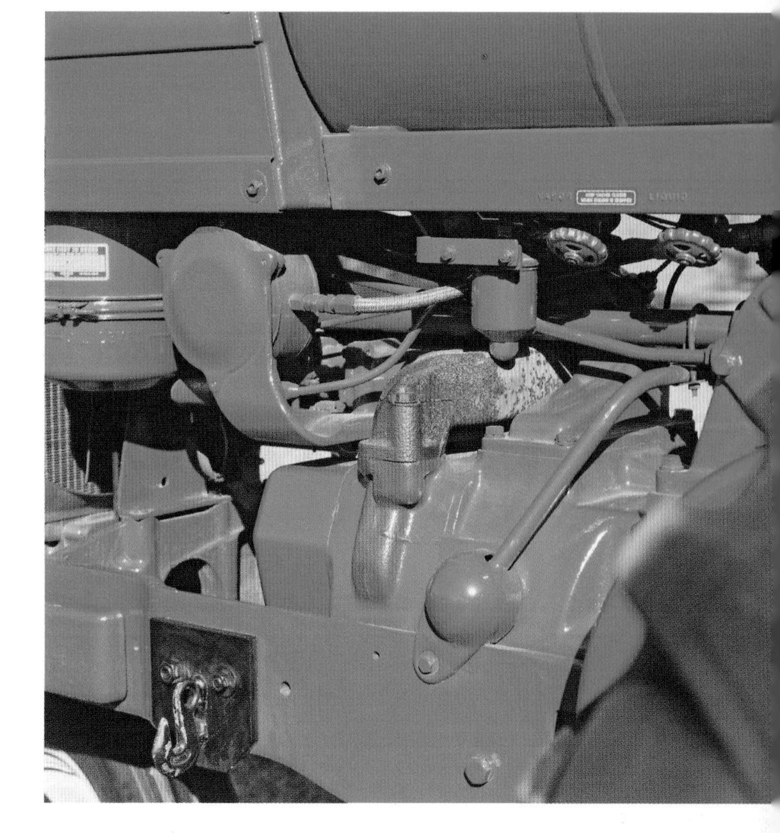

1953 John Deere 70 LP. The Model 70 LP row-crop was capable of handling the three-bottom 16-inch mounted plows. The fact that this particular one has a four-spoke steering wheel indicates that it was not equipped with power steering. The "Roll-O-Matic" tricycle front allowed the two wheels to react independently to uneven terrain. *Ralph W. Sanders*

1953 FORD NAA JUBILEE

1953 Ford NAA Jubilee. Although not providing the crop clearance of the tricycle row-crop tractors, the configuration of the Jubilee did allow cultivation of crops nearly as tall. Further, it could work in places with restricted overhead clearance such as in barns and orchards. *Ralph W. Sanders*

The introduction of the Ford-Ferguson 9N in 1939 was a momentous occasion in the history of tractor development. Every year following, the tractor got better as problems were corrected and improvements made. The 8N of 1948 was a big step improvement, but the NAA (1953), the last of the N-Series, was the best of the line and the "Ultimate N."

More than 840,000 of the N-type tractors were sold from 1939 to 1952 at a price starting at less than $600, eventually rising to about $1,200. At the peak, 9,000 tractors per month were being delivered (a good yearly volume for single models from other manufacturers). These tractors initiated transition from the row-crop configuration to the now standard-utility configuration, and it introduced the load-compensating three-point hitch, also now universally adopted. The Jubilee model is relatively rare with only about 78,000 being made in 1953.

Ford Motor Company celebrated its 50th anniversary in 1953. The 1953 NAA Jubilee, designed by young Henry Ford II's new management team, was the first completely new Ford tractor in 14 years. Prominently displayed above the grille was a circular emblem with the words "Golden Jubilee Model 1903–1953." *Golden* referred to the 50th anniversary; *Jubilee* is a biblical term for a 50-year period. The 1954 version of the tractor was essentially the same, and although generally called a *Jubilee*, the emblem has stars in place of the words.

The NAA was longer, taller, and 100 pounds heavier than the 1952 8N. It retained the utility configuration pioneered by the 9N and the paint scheme of the 8N but had an all-new 134-ci overhead valve four-cylinder engine. It also had an engine-driven hydraulic pump that avoided Ferguson's patent. The high-direct-low auxiliary gearbox was a popular option. This, along with the four-speed transmission, offered a top speed of 20 mph. The NAA weighs just under 3,000 pounds.

1957 FERGUSON 40/MASSEY-HARRIS 50

Two of the neatest small tractors of the time were the Ferguson 40 and the Massey-Harris 50. They were virtually identical under the skin and were the product of the merger of the two parent companies. And despite the conflict between the new joint company's competing dealers, the tractors turned out to be among the best.

The new company, Massey-Harris-Ferguson, was announced in late 1953. It was now the world's second-largest farm machinery company, behind International Harvester but ahead of Deere. It was agreed that a "two-line" policy would be maintained so that factories and dealerships would not be disrupted. The best-laid plans, however, do not always work out. The next three years were a mess for Massey-Harris-Ferguson (Massey-Ferguson by 1958). Sales, profits, and morale declined sharply. The goodwill of the merger soon deteriorated into open conflict between the "Reds" and the "Grays." And within the first year, Harry Ferguson left the company.

The policy emerged that the two-line approach would be continued but that each line would have to be strengthened so as to be competitive, not merely complementary. It was also decided that distinctions in styling would be maintained, even as the products became more similar.

Meanwhile, Ferguson engineers were completing work of the new Ferguson 35, an update of the Ferguson 30, to make it competitive with the Ford NAA. When Massey dealers saw what they were now up against from their own company, they squawked mightily. It was hurriedly decided to make a version of the 35 for Massey dealers, called the MH-50. The MH-50 was introduced in December 1955 and was produced to 1958.

Now Ferguson dealers were the ones to squawk. Therefore, in 1956, some new "Ferguson" sheet metal was added resulting in the tractor called the Ferguson 40. The color scheme was beige and metallic gray. The 50 and 40 were produced side by side until the end of the two-line policy. After that, the MH-50 alone was offered, but now it was called the MF-50. In addition to the 134-ci Continental four-cylinder engine, a three-cylinder 152-ci Perkins Diesel was also offered.

1957 Ferguson 40. Made to compete with the Ford Jubilee, the Ferguson 40 was about the same size and had the same engine displacement. Like the Jubilee, it was rated for three 14-inch or two 1-inch plows. The 40 was the last production tractor to bear only the Ferguson name. *Andrew Morland*

1957 PORSCHE 122

Ferdinand Porsche, the engineer behind the famous automobile company, had also been designing tractors in the 1930s. His son, Ferry Porsche, designed the postwar line of tractors in Germany. The burgeoning sports car business caused a licensing agreement to be made with Allgaier, also of Germany. Then, Mannesmann (MAN) purchased the rights in 1956 and continued production until 1963. Assets were then sold to Renault, and the Porsche name disappeared from tractors.

The 122 used a two-cylinder air-cooled diesel of 106 ci. It produced about 22 horsepower. A five-speed transmission gave a top speed of 13 mph. Weight of the 122 was about 3,000 pounds.

All of the Porsche tractors are rare. Not many have made their way to the United States.

1957 Porsche 122. Postwar production restrictions forced Porsche to license Allgaier for tractor production. The last two numbers of the numeric designator on all Porsche tractors indicates the nominal horsepower. Note the coil springs on the front axle. Suspended front ends were common on European tractors. *Ralph W. Sanders*

1958 JOHN DEERE 420 HC

The 420 series of Dubuque tractors came out a year before the other three-numbered tractors from Deere. These first 420s had all-green bodies, rather than the two-tone paint scheme that followed in the 1957 model year. Production continued through 1958 when the 30 Series came out. The 420 needed a bit more horsepower, so the bore was increased by 0.25 ci, upping the displacement to 113.3 ci. The compression ratio was also raised (on the gasoline engines) to 7.0:1. These changes gave about 20 percent more power. All-fuel, or distillate, versions of the 420 were offered along with the gasoline types, although they were not a significant proportion of the total with only about 1,000 being sold. In 1958, an LPG (Liquefied Petroleum Gas) option was offered. This engine used an 8.5:1 compression ratio to compensate for the lower BTU rating of LPG.

To provide adequate cooling capacity for the more powerful engines, the thermocycle, or gravity-type cooling system of the 40 was replaced by a water pump system with a pressurized radiator and a thermostat.

A variety of transmissions was offered for the 420. The High Crop version could have either a four- or five-speed type. Top speed for either was 12 mph.

For the 1958 model year, the vertical steering wheel was replaced by one mounted on an angle. This proved to be a much more comfortable position and more like that offered by the competition. This feature was not available, however, on the high-crop or LPG tractors. Tractors without power steering got a larger-diameter steering wheel.

The 420 H had a shipping weight of 3,400 pounds. Only a few more than 600 of the 420 H were built.

1958 John Deere 420 HC. The uncomfortable vertical steering wheel only was available on the 1958 HC version of the Model 420, and the "slant" steering wheel was not available on the LPG fuel version. *Ralph W. Sanders*

JOHN DEERE
Power Steering
420

1960 MINNEAPOLIS-MOLINE MOTRAC

Minneapolis-Moline came out with this handy crawler excavator in 1958, calling it the Gold Cat. For obvious reasons, it had trademark trouble with that, so they changed the name to Two Star, following the new numbering system for their wheeled tractors. Finally, in 1959 the MoTrac name was adopted.

The MoTrac used the same four-cylinder engine as the M-M Four Star wheeled tractor, which was a 206.5-ci overhead-valve type with the cylinders cast in pairs. Each pair could be separately removed from the rigid basepan for easy servicing. The engine produced around 45 horsepower. Versions for gasoline, diesel, or LPG fuel were available. The MoTrac was equipped with a five-speed transmission with a shuttle reverse.

Steering controls were by two levers curving up from either side of the cockpit to meet in the center. Each lever controlled first a clutch and then a brake for its corresponding track. There was a foot service brake and a foot master clutch pedal.

Fewer than 250 MoTracs were built, about two-thirds of which were powered by diesel engines.

1960 Minneapolis-Moline MoTrac. M-M's venture into the crawler-industrial market was characterized by its especially high bucket lift. *Andrew Morland*

1960 SAME 480 DTB ARIETE (RAM)

The SAME (Società Accomandita Motori Endotermici) company was founded in Treviglio (Bergamo), Italy, by Eugenio and Francesco Cassani in 1942, right in the middle of World War II, with bombardments, lack of raw materials, and very few production facilities. Nevertheless, the company established itself as a premier diesel engine tractor maker, the first outside Germany. It also pioneered the use of four-wheel drive for its tractors from as early as 1951.

The 480 DTB is a four-wheel-drive tractor with a four-cylinder, air-cooled diesel engine of 305 cid. It develops 80 horsepower on the PTO. The transmission has eight forward gears. A three-point hitch system is included with lower-link draft control.

Today, the company has morphed into SAME Deutz-Fahr (SDF), an Italy-based manufacturer of tractors, combine harvesters, other agricultural machines, engines, and equipment. It builds tractors under the SAME , Lamborghini, Hürlimann, and Deutz-Fahr brands, and combines under the Deutz-Fahr and Đuro Đaković brands.

SDF builds some tractors for AGCO, of which it owns a small percentage.

1960 SAME 480 DTB Ariete (RAM). This capable Italian tractor included all the modern features including the Automatic Control Unit that regulated plowing depth. *Andrew Morland*

1961 MASSEY-FERGUSON 85

First introduced in 1959, the big Massey-Ferguson 85 was the first and only five-plow tractor to be equipped with a draft-control three-point hitch. It was available in standard-tread, tricycle, and high-clearance versions, and with gasoline or LPG fuels. Some references mention a diesel version, but evidence indicates that likely none of those were made.

The engine used in the 85 was a Continental overhead-valve four-cylinder of 242 ci. An eight-speed, two-stick gearbox also provided two speeds in reverse. Top speed was about 20 mph. Weight was about 5,700 pounds for shipping and double that with ballast.

Although sold in limited numbers, the MF 85 was popular with farmers cultivating row crops, as the tractor could be equipped with a front-mounted cultivator system operated by the hydraulic control lever. Lights, deluxe seat, disk brakes, and power steering were standard equipment. The exhaust could be either vertical or run under the tractor with the muffler under the hood.

1961 Massey-Ferguson 85. A brute of a tractor, especially in this high-clearance version, the exhaust pipe reaches about eight feet above the ground. Instrumentation includes a tachometer, fuel gage, ammeter, oil pressure, and coolant temperature. The red panel light indicates when the live PTO is engaged. *Andrew Morland*

1961 COCKSHUTT 580 STANDARD

In 1958, Cockshutt introduced a complete new line of tractors at the same time: the 540, 550, 560, and 570. Raymond Loewy, an automobile designer of the era, designed the sheet metal of the 500 series. The design set a new standard in modern styling. The 540 was powered by a 30-horsepower Continental gas engine of 162 ci (2- to 3-plow rating); the 550 was powered by a 40-horsepower Hercules 198-ci gas or diesel (3-plow rating); the 560 was powered by a 50-horsepower Perkins diesel of 270 ci (4-plow rating); and the 570 was powered by a 65-horsepower Hercules diesel of 298 ci (5-plow rating). The big-brother 580 was never mass produced; the first three hand-assembled units were on the shop floor in the plant when the shutdown order came in early 1962. It was a 100-horsepower unit, and only one tractor escaped demolition: it is in the hands of a collector.

The 580 was powered by a six-cylinder Perkins diesel of 354 ci. The standard six-speed transmission was augmented by a two-speed rear end arrangement, which provided twelve speeds forward and three in reverse.

All of the Cockshutt models of this period were state-of-the-art tractors, but due to termination of the Canadian company (taken over by White in 1962), they were built in limited numbers.

1961 Cockshutt 580 Standard. Shades of what might have been! The Cockshutt 580 died a'bornin' like the Ferguson Model 60, victim of consolidating companies and rampant competition. *Ralph W. Sanders*

580 SUPER
COCKSHUTT
DIESEL

1961 FORD 6000

The red-and-gray Ford 6000 was a completely new tractor with all the modern bells and whistles, including a new 10-speed power-shift transmission. It seems the tractor was rushed into the hands of farmers without sufficient testing and against the advice of senior engineers. Most were recalled in 1964, slightly restyled, their shortcomings corrected, and their paint scheme changed to blue and white.

These, for 1965, with a new two-piece grille, were now labeled the Commander 6000. Customers had a choice of row-crop or standard-tread versions. The Commander 6000 was much the same as before, employing a number of interesting features that made it competitive with others in its size class. These included a Category 2 three-point hitch employing lower-link draft sensing, a hydraulic system that used an accumulator (to store hydraulic power to supplement the pump during periods of high demand), standard power disk brakes and power steering, operator-sitting or -standing steering wheel positions, and a unique PTO. This PTO provided 540 or 1,000 rpm outputs (operator's choice) at either of two engine speeds: 2,225 rpm for high-horsepower applications or 1,730 rpm for more moderate conditions.

The 10-speed power shift transmission was retained and was the only offering. The diesel engine displaced 242 ci, and the gasoline and LPG versions displaced 223 ci. All were in the 60- to 65-horsepower class and weighed about 6,500 pounds (shipping). The Commander 6000, like the 6000 before it, was unique among Ford Tractors in that it was built on a frame rather than on the traditional unit-construction concept.

All were quite limited in production, but the original red and white 6000 is rare.

1961 Ford Model 6000. Because of myriad troubles plaguing the original 6000s, the "New-Blue" improved 6000 was introduced in 1962. At the same time, it was announced that Ford would recall the old 6000 tractors. Every owner of a red and gray 6000 would be given a new blue and gray one with no other stipulations than they must put their old tires and rims on the new tractor. The red and gray tractors were to be either returned to Detroit for rebuild, or dealers could buy a kit and do the rebuilds locally. The one shown must have escaped the recall. *Andrew Morland*

1962 DOE TRIPLE D

If you wanted high horsepower and traction in the mid-1950s, you pretty much had to go with a crawler. If you couldn't put up with the inconvenience of a crawler, you could buy a Doe Triple D.

The Doe Triple Ds were made up of two Fordson Super Majors hooked in tandem (front wheels removed from both). The hydraulic power from the front unit was used for articulated steering and for actuating the controls (gearshift levers and "throttle"). Rear hydraulics powered the three-point hitch. At least on the earliest production models, the front engine had to be started before the operator mounted the rear unit, but that problem was later overcome. Total horsepower was a respectable 96 on the drawbar. Of course, belt or PTO horsepower only reflected the power of one engine. Gross weight was about 12,000 pounds plus ballast. Top speed was 15 mph.

More than 300 of these were sold by Doe, mostly in England. When Ford stopped building the Super Major in 1964, Doe made some similar machines using Ford 5000 units, but by then production four-wheel drives were becoming commonplace and the market for these extra-long machines dried up.

1962 Doe Triple D. The twin Super Major Doe Triple D tractor was capable of pulling an eight-bottom plow. Driven from the rear unit, this monster was over 23 feet long. *Andrew Morland*

1966 CASE 1200 TRACTION KING

Case's first four-wheel-drive tractor, the 10-ton Traction King (TK) 1200, was also Case's first turbocharged tractor. It was the first tractor tested at the University of Nebraska with four-wheel power hydraulic brakes. The 1200 TK was not articulated but employed four-wheel steering. The front wheels were turned by the steering wheel with normal hydrostatic power steering. The rear wheels used a separate hydraulic system and were controlled independently by a hand lever.

The engine used in the 1200 was a 451-ci six-cylinder turbocharged Case-built diesel using the Lanova "Powrcel" induction. It produced 120 horsepower at the PTO. A category III three-point hitch was used with lower-link draft sensing.

The 1200 was in production from 1964 to 1969 when it was replaced by the 1470 Traction King.

1966 Case 1200 Traction King. This tractor had a bare weight of almost 17,000 pounds. Its massive 28.1–26 tires applied over 100 drawbar horsepower to the ground. Because of the unique two-control steering, crab steering was possible. The one shown has the optional cab. *Ralph W. Sanders*

1973 FORD 1000

In 1973, Ford announced an agreement with the Japanese Ishikawajima-Shibaura Machinery Company to build a small tractor according to Ford's specifications and following Ford's styling patterns. The new "compact" was to carry on for the famous N-Series Ford tractors for gardeners and small-acreage farmers.

The 1000 used a two-cylinder diesel of 77.6 ci, which put out about 23 horsepower at 2,500 rpm. The basic weight of the tractor was 2,350 pounds. Two shift levers controlled nine forward speeds and three reverse. Top speed was 8 mph. The Category 1 three-point hitch was live, but the 540 rpm PTO was not. Standard equipment included a tachometer, a differential lock, a ROPS, and a deluxe seat. A factory cab was an option. The 1000 was produced for Ford from 1973 to 1976.

1973 Ford 1000. Besides the nifty cab, options for the 1000 included the front lift, front weights, front and rear wheel weights, a roll bar, and a seat belt. It was replaced in 1975 by the Model 1600, which was restyled but otherwise much the same. *Andrew Morland*

FORD 1000
FORD

1976 CASE 1570 SPIRIT OF '76

During the first half of the 1976 model year, Case offered the 1570 with a special red-white-and-blue, stars-and-stripes paint scheme to commemorate the country's bicentennial. The 1570, built from 1976 to 1978, was a 180 PTO-horsepower, two-wheel-drive tractor with a standard air-conditioned cab. A special deluxe swivel seat was featured. A 504-ci direct-injection six-cylinder turbo diesel engine was used along with a 12f/3r transmission with partial-range power shifting. Top speed was 20 mph. Weight (for shipping) was 16,000 pounds.

Since the Spirit of '76 versions only represent a half-year's limited production as dealer promotional specials, they are rare. In addition, many of the original paint jobs are now in disrepair.

1976 Case 1570 Spirit of '76. In the 1970s, powerful two-wheel drive tractors were popular. This Case 1570, with 180 horsepower, is a good example. The patriotic paint job was available only in the first half of 1976. *Ralph W. Sanders*

1976 MERCEDES-BENZ UNIMOG/CASE M-B 4/94

Since the early 1950s, Mercedes-Benz had been marketing this truck/tractor combination. There have been a variety of sizes and horsepower but the same recognizable shape. Called the Unimog, it is basically a four-wheel-drive utility vehicle with a truck bed. It is capable of hauling produce to town or plowing a field. Unimog is a contraction of the German words *Universal-Motor-Gerät,* meaning "Universal Motor Machine."

A version called the 30 was tested at the University of Nebraska in 1957. It had a four-cylinder diesel of 108 cid, which put out 27 horsepower at 2,550 rpm. In 1976, Case listed one in their catalog as the Case M-B 4/94. This one was of the same basic configuration but sported a six-cylinder diesel of 346 ci, producing 70 horsepower at 2,550 rpm.

Unimogs have high ground clearance, because the axles employ portal, or bull, gears at each end. They can be equipped with three-point hitches and PTOs. The Case version weighed 8,500 pounds and had a road top speed of 30 mph.

1976 Mercedes-Benz Unimog/Case M-B 4/94. Unimogs come in a variety of sizes but in generally the same configuration. The one shown is set up for mounting a snow plow. Plowing deep winter snow is a task for which they are well suited. *Andrew Morland*

1983 KUBOTA M7950 DT

Built from 1983 to 1991, the 75 PTO-horsepower Kubota evolved slowly before morphing into the M8580. The four-cylinder Kubota, naturally aspirated diesel engine displaced 262 ci. The tractor was equipped with four-wheel drive and had an optional air-conditioned cab. A twelve-speed fixed-ratio selective-gear transmission offered three speeds in reverse. Top speed in road gear was 18 mph. Basic weight was 7,100 pounds.

1983 Kubota M7950 DT. Equipped with front-end weights and a live three-point hitch and PTO in the back, the M7950DT was a thoroughly modern early 1980s tractor. The 12-speed transmission required two shift levers. *Andrew Morland*

1988 FORD VERSATILE BI-DIRECTIONAL 256

Talk about a neat concept; the bi-directional Fords certainly looked versatile. You could have a three-point hitch and PTO on both ends of the tractor, a swivel-seat, and a shuttle shift. In Wisconsin, for example, you could have a mower on the one end and a snow blower on the other and be ready whatever the weather. But high cost kept sales numbers low.

Ford-New Holland acquired the Bi-Directional with the acquisition of the Versatile Farm Equipment Company of Winnipeg, Manitoba, Canada, in 1987. Production was continued with minor refinements and a Ford oval above the grille. Originally, a CDC diesel engine was used. It was a four-cylinder turbocharged unit of 239 ci. Rated at 2,500 rpm, it produced about 85 horsepower. A hydrostatic continuously variable transmission was provided, with a three-speed fixed-ratio auxiliary. Top speed was 20 mph. Steering was by hydraulic articulation. The tractor weighed in at 9,000 pounds for its Nebraska Test (No. 1,518).

In 1989, Ford substituted its own turbo-charged diesel for the CDC (Consolidated Diesel Corporation) engine used previously, and Ford colors were applied. In 1990, the Bi-Directional was restyled to be more like others in the Ford-New Holland line, and the number was changed to 9030, also to be more like the others. Then, in 1995, the Ford name disappeared from tractors after Fiat bought New Holland.

1988 Ford Versatile Bi-Directional 256. Proving the versatility of the concept, the driver is facing comfortably forward while back-blading the driveway with the loader bucket. *Andrew Morland*

VERSATILE

1997 JCB FASTRAC

Relatively few of these Fastracs have been imported into the United States from Great Britain. The JCB Fastrac line of tractors defines modern versatility. Most, if not all, are working tractors, but they certainly are interesting to collectors. They are built by the J. C. Brandford Group and distributed worldwide. Remarkable features start with a full four-wheel suspension system and gearing that takes the tractor to 40 mph. If, however, the model has full hydrostatic steering (no mechanical connection), laws may require speed to be limited to 30 mph. All models are equipped with disk air brakes on all four wheels. All have selectable four-wheel drive and four-wheel steering (below 12 mph). All have center differentials that allow sharp turns in four-wheel drive on hard surfaces. All have two-speed PTOs and three-point hitches front and back. Earlier versions used Perkins diesels exclusively but later ones will also use Cummins engines. Earlier Fastracs had fixed-ratio 18f-6r gearboxes. Later versions had a variety of power shift and continuously variable transmissions (CVT). Horsepower ranged between 135 and 300. Weights start at about 14,000 pounds.

1997 JCB Fastrac. With front and back three-point and PTO-powered haying equipment, this JCB Fastrac demonstrates modern productivity. *Andrew Morland*

N665 GEH
JCB
LELY
LELY OPTIMO 280 C

CREAM of the Crops

THE FOLLOWING ARE TRACTORS that would rate very high on the desirability lists of most every serious tractor person. In some cases, there were a lot of these manufactured, but time has taken a toll on their availability. For other models, fewer than 100 are out there, and very few ever become available for purchase. The hope of acquiring any one of these usually goes beyond the luck of the hunt. You might find a collector with two who is willing to sell one, or you might have to take the least-desirable option: wait for an estate sale.

1918 WATERLOO BOY R

The tractor that got John Deere into the tractor business took shape back in 1915 under the auspices of the Waterloo Gasoline Engine Company of Waterloo, Iowa. The name "Waterloo Boy" is thought to be both a reference to the name of the town and to the welcome "water boy," whose job it was to carry cool, refreshing water to the thirsty members of threshing crews.

The Waterloo Gasoline Engine Company had made tractors in several styles prior to 1915 when the R was settled upon. It was produced through 1919, overlapping its successor, the N, which came out in 1917. Deere and Company bought the Waterloo Gasoline Engine Company in 1918. Production of the N continued through 1924, overlapping its successor, the John Deere D.

There were more than 10,000 Waterloo Boy Rs made, of which some 4,000 were sent to Britain during World War I. Imported to alleviate the food shortage resulting from the German submarine threat, these were renamed "Overtime" by their importer.

The Waterloo Boy R was a four wheel, rear-wheel-drive machine with one speed forward and one in reverse. The engine was a horizontal side-by-side two-cylinder of 395 cid; this type of engine would characterize Deere tractors for the next 42 years. Displacement was increased to 465 ci in 1917. Steering was by swing axle, chain, and bolster. The steering wheel was on the right, the gas tank in the front. The cooling radiator was generally on the right. A belt-driven fan induced the cooling. The R weighed in at about 5,900 pounds. Only the N was tested at the University of Nebraska but was generally in the 20- to 25-belt-horsepower class.

1918 Waterloo Boy R. The most visible difference between the Waterloo Boy Model R and N is the R has a much smaller diameter ring drive gear in each rear wheel. *Andrew Morland*

1925 JOHN DEERE D (SPOKER)

When John Deere took over the Waterloo Boy outfit in 1918, work was already underway on a new, more modern tractor design to replace the Waterloo Boy. John Deere engineers continued the development through four prototypes, each given a letter designator A through D. The D version became the John Deere D. It would be one of the longest-produced US tractor models, running from 1923 to 1953 with around 160,000 manufactured. However, not many were made with spoked versus solid flywheels, which makes the "Spoker" rare and valuable.

Serial numbers 30401 through 31279, made in 1923–24, were furnished with 26-inch spoked flywheels. Serial numbers 31280 through 36248 had 24-inch spoked flywheels. This means that there were about 5,800 made (Model D serial numbering became confused when continued Waterloo Boy production serial numbers began to overlap those of the D).

The two-cylinder side-by-side horizontal engine of the D was largely the same as that of the Waterloo Boy but larger in displacement at 465 ci. A two-speed transmission was provided a top speed of 3.25 mph. During University of Nebraska Test No. 102, performed in 1924, the D recorded 30.4 belt horsepower. Weight was 4,100 pounds.

1925 John Deere D (Spoker). The 26-inch spoked flywheel was rolled over by hand for engine starting. *Ralph W. Sanders*

A characteristic of the 1924 D was a left-hand steering wheel. *Ralph W. Sanders*

The D was built on the unit-frame concept and featured a roller-chain final drive system. *Ralph W. Sanders*

JOHN DEERE
JOHN DEERE

1930 MASSEY-HARRIS GP 15/22

Massey-Harris was somewhat slow getting into the tractor business, first taking on lines developed by others. After World War I, Massey-Harris acquired the Case Plow Works and its Racine, Wisconsin, manufacturing facility. Also with the deal, Massey-Harris got the rights to Case's Wallis tractor line. Its engineering department directly undertook the development of Massey's first in-house tractor design: the GP 15/22.

The GP (General Purpose) was an all-new, radical four-wheel-drive machine powered by an L-head 226-ci Hercules four-cylinder inline vertical engine that could be configured for either gasoline or kerosene fuel. It had four equal-size wheels powered through a three-speed transmission, a transfer case, differentials on each axle, and lastly, through final drive meshes with small gears on the ends of each axle that mated with large diameter gears on each wheel. Since this final mesh was at the top of the wheel, ample crop clearance was provided under the axles for cultivation of taller crops.

The GP's front wheels were steerable through universal joints on the drive shafts, and there were individual brakes on each front wheel to aid in steering. The rear axle was free to swivel around the differential input so that all four wheels could remain in contact with the ground on uneven surfaces.

The GP was available with an electrical system, starter, and lights. Also offered was an implement-lift system. It came in four different tread widths for row-spacing, though these were fixed and not adjustable. Orchard fenders were available for the narrowest version. Railroad, golf course, and industrial versions were offered.

Production of the GP 15/22 began in 1930. In 1936, an improved version was introduced, identified by a down-sloping hood. The engine, now with overhead valves, was the same displacement and power. The GP name was dropped, and it was simply called the 4-WD. Production seems to have ended in 1936 with only some 3,000 of all types having been built.

1930 Massey-Harris GP 15/22. The M-H GP sales were disappointing, and so was its performance. Its main competition, the Fordson, pulled 3,300 pounds against the GP's maximum pull of 3,200 pounds, but the slippage for the Fordson was almost 20 percent, versus 8 percent for the GP. *Ralph W. Sanders*

1936 JOHN DEERE BW-40

John Deere, ever willing to please customers, started making variations to its B tractor line almost as soon as production started in 1935. The first modification to the conventional GP (General Purpose) B was the BN, with a single front wheel but otherwise the same. It was made for California vegetable growers and also became known as the B Garden Tractor. The next variation was the BW with an adjustable-width front end (rear-wheel spacing was adjustable on all B GPs).

An odd variation of the BW was the super-rare BW-40 or "Special Narrow" BW. Special axles were used front and rear, to allow tread widths as narrow as 40 inches. Maximum tread width was 72 inches, but that required extensions for the front axles. It seems that only six of these were manufactured in 1935 and 1936.

Power for the BW-40 came from the standard two-cylinder engine of 149 ci. A four-speed transmission was provided. Shipping weight was 2,700 pounds.

1936 John Deere BW-40. Only a few of these tractors were made in 1935 and 1936 for farmers that needed 40-inch wheel spacing (minimum). Serial number tags are found on the transmission housing just ahead of the flywheel. *Ralph W. Sanders*

1938 MINNEAPOLIS-MOLINE UDLX COMFORTRACTOR

The "Holy Grail" of tractor collecting, the UDLX is one of the most sought-after tractors of all time. The best estimates are that 150 were made between 1938 and 1941. Very few were delivered to farmers, who worked them in their fields by day and then drove them to town in the evening. Most were driven by custom threshermen able to scoot between jobs towing the thresher at 40 mph. Stories recount Minneapolis-Moline salesmen driving Comfortractors to visit dealerships; that, however, seems to have been rare indeed.

The UDLX (or U-Deluxe) Comfortractor was a version of the M-M U Series tractors. The UDLX featured items like a shift-on-the-fly five-speed transmission, tip-out windshields, windshield wipers, high- and low-beam headlights, taillights, heater, speedometer, and even a cigar lighter. There was somewhat cramped seating for three.

The fully enclosed cab was comfortable for road trips, but since there were no hydraulics, the back door had to be open for access to trailed implements. Also, there were no provisions for a belt pulley or PTO shaft, so the tractor was only useful for pulling jobs. Since there were no springs on either axle, the UDLX tended to waddle down the road.

Power for the UDLX came from an overhead-valve four-cylinder gasoline engine of 284 ci. It produced about 42 horsepower. The tractor weighed 4,500 pounds.

While most of these have already undergone extensive restoration, those who have done so will say that it was not an inexpensive job. One of the main difficulties is that the cab structure was made of wood. The years have taken their toll, producing dry rot in these structure members, which now have to be carefully copied and remade.

1938 Minneapolis-Moline UDLX Comfortractor. This tractor featured a very stylish front end for the times, but the front axle did not have springs, which made for a bone-shaking ride on dirt farm roads. *Ralph W. Sanders*

The fold-up seat allowed occupants to enter the cab. Access was still restricted by the transmission housing. *Ralph W. Sanders*

The stylish lines compared favorably to the finest 1938 automobiles. Note the radio antenna on the left windshield frame and the spotlight on the roof. *Ralph W. Sanders*

1938 GRAHAM-BRADLEY 503-103

Possibly the most stylish tractor of all time, the Graham-Bradley was particularly striking in 1938. Graham-Paige Motors Corporation of Detroit, Michigan, produced this nifty tractor exclusively for the giant mail-order firm of Sears, Roebuck & Company of Chicago. The tractor was called the Graham-Bradley as the Bradley name had been used by Sears for farm items for some time.

Graham-Paige, an automobile manufacturer, had its heritage in the Paige-Detroit Motor Car Company, founded in 1908. In 1928, the Graham brothers acquired the company, renaming it Graham-Paige Motors Corporation, and it continued producing stylish automobiles until World War II.

The 503-103 tricycle version was offered through Sears from 1938 to 1941. A wide-front version, the 503-104, was introduced in 1939 and was carried to 1941, as well. Both were otherwise the same, using Graham-Page's 218-ci L-head six-cylinder engine from its car but governed to 1,500 rpm. As such it produced a maximum belt horsepower of 30. The belt pulley was downstream of the four-speed transmission and therefore had four belt-pulley speeds. Top speed of the Graham-Bradley was a racy (for 1938) 20 mph.

A clear picture of how many Graham-Bradley tractors were made is not apparent, but Graham car production in those days only amounted to 10,000 to 12,000 automobiles. Drawing a parallel to the tractors means the Graham-Bradleys are likely quite rare.

At the end of World War II, Graham-Paige president Joseph Frazer merged his company with that of Henry J. Kaiser, the ship-building magnate. The resulting Kaiser-Frazer automobile company never revived tractor production.

1938 Graham-Bradley 503-103. The Graham-Bradley tractors featured some unlikely attributes for their time, besides the striking styling. Standard items included a comfortable seat with a back, an under-hood muffler, a Delco-Remy 6-volt electrical system with distributor ignition, and a transmission that gave governed speeds of 2 mph to 20 mph. *Ralph W. Sanders*

GRAHAM-BRADLEY

1950 FUNK-FORD 8N V-8

Tractor people familiar with aircraft know that the Stearman biplane with the 450-horsepower Pratt & Whitney engine in place of the original 225-horsepower Continental is known as the "Bull Stearman." Well, the Funk-Ford with the 239-ci V-8 engine conversion could be known as the "Bull Ford." Even governed to 2,400 rpm, the Ford industrial V-8 more than tripled the horsepower of the 8N.

Most Funk-Fords were converted by Ford tractor dealers from new 8Ns, but kits were also available for 9Ns, 2Ns, and later Jubilees. Besides the V-8s, Ford industrial six-cylinder L-head engines were more commonly used for conversions. In 1953, Jubilee Fords were converted to the overhead-valve six. With any of these engines, the Funk-Fords became the most powerful wheel tractors available at that time.

Just before World War II started, the Funk brothers' Akron, Ohio, airplane-manufacturing company went into receivership. They had built a two-seat lightplane powered by a Ford four-cylinder Model B car engine. By 1940, the availability of those engines had completely dried up. The brothers switched to a Lycoming engine but nevertheless went bankrupt. Their receivership was picked up by a foundryman in Coffeeville, Kansas, and airplane production continued along with the foundry business. While traveling on sales calls, one of the Funk brothers happened to stop at a Ford dealership where he found a 9N Ford tractor that had a Ford six crudely installed. He offered to make proper adapter castings and other necessary parts for conversion kits. The rest, as they say, is history.

The V-8 powered 8N with dual vertical exhausts gives unbeatable stereo music while going where no other 8Ns can go. There is some risk of overtorquing the powertrain, but unless you get stuck, that is generally not a problem. Where Funk-Fords really shine is when a lot of torque is taken out

1950 Funk-Ford 8N V-8. When viewed from a distance, this appears to be an ordinary 8N, but up close, the higher hood can be seen, which allows clearance for the air cleaner and carburetor. Also, the radiator is much larger and taller. Finally, most of these have the largest back tires that will fit. *Andrew Morland*

through the PTO, such as in mowing jobs. For best results, a high/low auxiliary gearbox is used, which was an option of later 8Ns.

After only about 100 V-8 conversions were made, the Funk factory burned. The factory was rebuilt, but the production of aircraft and tractor conversions never resumed. Deere & Company eventually bought the factory, and it is now producing castings for Big Green.

L-head six-cylinder conversions are interesting but much more common than the V-8. Overhead valve sixes are about as rare as the V-8.

1960 MASSEY-FERGUSON 98/OLIVER SUPER 99 GM

The Oliver Super 99 GM was built between 1955 and 1959. For 1959 and 1960, some 500 of them were sold to Massey-Ferguson to be rebadged and resold as Massey-Ferguson 98s (along with a Massey grille in place of the Oliver grille). Updated Oliver versions, labeled 990 and 995, continued in production through 1961; some of these, and later Super 99GMs, were equipped with torque converters. All of these were truly awesome machines, but the Massey-Ferguson version, with only 500 being sold, is truly rare.

Probably the most interesting thing about this tractor is its three-cylinder, two-cycle diesel, made by General Motors. This engine, designated by GM as the 3-71 (for 3 cylinders of 71 cubic inches each), had a total displacement of only 213 ci, but because each cylinder had a power-stroke on each revolution, it had power as if it was twice as big, and at its rated speed of 1,675 rpm, it sounded like it was going twice as fast. In addition, the GM engine was blower-scavenged, which meant it used a Rootes-type blower to purge exhaust gases and to supercharge the compression. No intake valves were used, but the pistons uncovered intake ports when at the bottom of their strokes. When these ports were uncovered, conventional exhaust valves opened, and the blower swept exhaust gases out. The blower also added to the sound, so that one of these in full flight howled like a banshee. A characteristic of the engine was that it lost power quickly if the rpm dropped. Therefore, unlike other diesels, it was best to keep it howling at all times.

Most of these were built in the South Bend, Indiana, plant, but production after 1958 was transferred to the Charles City, Iowa, plant. At 85 horsepower, these were the most powerful wheel tractors of their day. Fully ballasted, they would weigh in at more than 15,000 pounds.

1960 Massey-Ferguson 98. This is what you needed in 1960 if you had a lot of acreage to plow. These monsters (or the equivalent Oliver Super 99 GM) could haul six 16-inch plows fast enough to plow a "forty" in a day. Hopefully you had ear protection! *Andrew Morland*

Index

About the Author

Robert N. Pripps has authored and co-authored dozens of farm tractor books, including *Classic Farm Tractors, Vintage Ford Tractors, Big Book of Caterpillar, Big Book of Massey,* and more. Pripps lives near Park Falls, Wisconsin, where he owns a maple syrup farm.